European Federation of Corrosion Publications
NUMBER 61

Inter-laboratory study on electrochemical methods for the characterisation of CoCrMo biomedical alloys in simulated body fluids

Edited by
A. Igual Munoz & S. Mischler

EFC

EUROPEAN FEDERATION OF CORROSION
FÉDÉRATION EUROPÉENNE DE LA CORROSION
EUROPÄISCHE FÖDERATION KORROSION

Published for the European Federation of Corrosion
by CRC Press
on behalf of
The Institute of Materials, Minerals & Mining

CRC Press
Taylor & Francis Group
Boca Raton London New York

CRC Press is an imprint of the
Taylor & Francis Group, an informa business

IOM3
The Institute of Materials, Minerals and Mining

CRC Press
Taylor & Francis Group
6000 Broken Sound Parkway NW, Suite 300
Boca Raton, FL 33487-2742

First issued in paperback 2020

CRC Press is an imprint of the Taylor & Francis Group, an Informa business

ISBN-13: 978-1-907625-00-8 (pbk)
ISSN 1354-5116

Visit the Taylor & Francis Web site at
http://www.taylorandfrancis.com

and the CRC Press Web site at
http://www.crcpress.com

Typeset by the Charlesworth Group, Wakefield, UK.

Contents

European Federation of Corrosion (EFC) publications: Series introduction

The European Federation of Corrosion (EFC), incorporated in Belgium, was founded in 1955 with the purpose of promoting European cooperation in the fields of research into corrosion and corrosion prevention.

Membership of the EFC is based upon participation by corrosion societies and committees in technical Working Parties. Member societies appoint delegates to Working Parties, whose membership is expanded by personal corresponding membership.

The activities of the Working Parties cover corrosion topics associated with inhibition, cathodic protection, education, reinforcement in concrete, microbial effects, hot gases and combustion products, environment-sensitive fracture, marine environments, refineries, surface science, physico-chemical methods of measurement, the nuclear industry, the automotive industry, the water industry, coatings, polymer materials, tribo-corrosion, archaeological objects and the oil and gas industry. Working Parties and Task Forces on other topics are established as required.

The Working Parties function in various ways, e.g. by preparing reports, organising symposia, conducting intensive courses and producing instructional material, including films. The activities of Working Parties are coordinated, through a Science and Technology Advisory Committee, by the Scientific Secretary. The administration of the EFC is handled by three Secretariats: DECHEMA e.V. in Germany, the Fédération Française pour les sciences de la Chimie (formely Société de Chimie Industrielle) in France, and The Institute of Materials, Minerals and Mining in the UK. These three Secretariats meet at the Board of Administrators of the EFC. There is an annual General Assembly at which delegates from all member societies meet to determine and approve EFC policy. News of EFC activities, forthcoming conferences, courses, etc., is published in a range of accredited corrosion and certain other journals throughout Europe. More detailed descriptions of activities are given in a Newsletter prepared by the Scientific Secretary.

The output of the EFC takes various forms. Papers on particular topics, e.g. reviews or results of experimental work, may be published in scientific and technical journals in one or more countries in Europe. Conference proceedings are often published by the organisation responsible for the conference.

In 1987 the, then, Institute of Metals was appointed as the official EFC publisher. Although the arrangement is non-exclusive and other routes for publication are still available, it is expected that the Working Parties of the EFC will use The Institute of Materials, Minerals and Mining for publication of reports, proceedings, etc., wherever possible.

The name of The Institute of Metals was changed to The Institute of Materials (IoM) on 1 January 1992 and to The Institute of Materials, Minerals and Mining with effect from 26 June 2002. The series is now published by Maney Publishing on behalf of The Institute of Materials, Minerals and Mining.

P. McIntyre
EFC Series Editor
The Institute of Materials, Minerals and Mining, London, UK

EFC Secretariats are located at:

Dr B. A. Rickinson
European Federation of Corrosion, The Institute of Materials, Minerals and Mining, 1 Carlton House Terrace, London SW1Y 5AF, UK

Mr M. Roche
Fédération Européenne de la Corrosion, Fédération Française pour les sciences de la Chimie, 28 rue Saint-Dominique, F-75007 Paris, France

Dr W. Meier
Europäische Föderation Korrosion, DECHEMA e.V., Theodor-Heuss-Allee 25, D-60486 Frankfurt-am-Main, Germany

Volumes in the EFC series

* indicates volume out of print

1 **Corrosion in the nuclear industry**
*Prepared by Working Party 4 on Nuclear Corrosion**

2 **Practical corrosion principles**
*Prepared by Working Party 7 on Corrosion Education**

3 **General guidelines for corrosion testing of materials for marine applications**
*Prepared by Working Party 9 on Marine Corrosion**

4 **Guidelines on electrochemical corrosion measurements**
*Prepared by Working Party 8 on Physico-Chemical Methods of Corrosion Testing**

5 **Illustrated case histories of marine corrosion**
Prepared by Working Party 9 on Marine Corrosion

6 **Corrosion education manual**
Prepared by Working Party 7 on Corrosion Education

7 **Corrosion problems related to nuclear waste disposal**
Prepared by Working Party 4 on Nuclear Corrosion

8 **Microbial corrosion**
*Prepared by Working Party 10 on Microbial Corrosion**

9 **Microbiological degradation of materials and methods of protection**
Prepared by Working Party 10 on Microbial Corrosion

10 **Marine corrosion of stainless steels: chlorination and microbial effects**
Prepared by Working Party 9 on Marine Corrosion

11 **Corrosion inhibitors**
*Prepared by the Working Party on Inhibitors**

12 **Modifications of passive films**
*Prepared by Working Party 6 on Surface Science**

13 **Predicting CO_2 corrosion in the oil and gas industry**
*Prepared by Working Party 13 on Corrosion in Oil and Gas Production**

14 **Guidelines for methods of testing and research in high temperature corrosion**
Prepared by Working Party 3 on Corrosion by Hot Gases and Combustion Products

15 **Microbial corrosion: Proceedings of the 3rd International EFC Workshop**
Prepared by Working Party 10 on Microbial Corrosion

16 **Guidelines on materials requirements for carbon and low alloy steels for H_2S-containing environments in oil and gas production (3rd Edition)**
Prepared by Working Party 13 on Corrosion in Oil and Gas Production

17 **Corrosion resistant alloys for oil and gas production: guidance on general requirements and test methods for H_2S service (2nd Edition)**
Prepared by Working Party 13 on Corrosion in Oil and Gas Production

18 **Stainless steel in concrete: state of the art report**
*Prepared by Working Party 11 on Corrosion of Steel in Concrete**

19 **Sea water corrosion of stainless steels: mechanisms and experiences**
Prepared by Working Party 9 on Marine Corrosion and Working Party 10 on Microbial Corrosion

20 **Organic and inorganic coatings for corrosion prevention: research and experiences**
Papers from EUROCORR '96

21 **Corrosion-deformation interactions**
CDI '96 in conjunction with EUROCORR '96

22 **Aspects of microbially induced corrosion**
Papers from EUROCORR '96 and EFC Working Party 10 on Microbial Corrosion

23 **CO_2 corrosion control in oil and gas production: design considerations**
*Prepared by Working Party 13 on Corrosion in Oil and Gas Production**

24 **Electrochemical rehabilitation methods for reinforced concrete structures: a state of the art report**
Prepared by Working Party 11 on Corrosion of Steel in Concrete

25 **Corrosion of reinforcement in concrete: monitoring, prevention and rehabilitation**
Papers from EUROCORR '97

26 **Advances in corrosion control and materials in oil and gas production**
Papers from EUROCORR '97 and EUROCORR '98

27 **Cyclic oxidation of high temperature materials**
Proceedings of an EFC Workshop, Frankfurt/Main, 1999

28 **Electrochemical approach to selected corrosion and corrosion control**
Papers from the 50th ISE Meeting, Pavia, 1999

29 **Microbial corrosion: proceedings of the 4th International EFC Workshop**
Prepared by the Working Party on Microbial Corrosion

30 **Survey of literature on crevice corrosion (1979–1998): mechanisms, test methods and results, practical experience, protective measures and monitoring**
Prepared by F. P. Ijsseling and Working Party 9 on Marine Corrosion

31 **Corrosion of reinforcement in concrete: corrosion mechanisms and corrosion protection**
Papers from EUROCORR '99 and Working Party 11 on Corrosion of Steel in Concrete

32 **Guidelines for the compilation of corrosion cost data and for the calculation of the life cycle cost of corrosion: a working party report**
Prepared by Working Party 13 on Corrosion in Oil and Gas Production

33 **Marine corrosion of stainless steels: testing, selection, experience, protection and monitoring**
Edited by D. Féron on behalf of Working Party 9 on Marine Corrosion

34 **Lifetime modelling of high temperature corrosion processes**
Proceedings of an EFC Workshop 2001 Edited by M. Schütze, W. J. Quadakkers and J. R. Nicholls

35 **Corrosion inhibitors for steel in concrete**
Prepared by B. Elsener with support from a Task Group of Working Party 11 on Corrosion of Steel in Concrete

36 **Prediction of long term corrosion behaviour in nuclear waste systems**
Edited by D. Féron on behalf of Working Party 4 on Nuclear Corrosion

37 **Test methods for assessing the susceptibility of prestressing steels to hydrogen induced stress corrosion cracking**
By B. Isecke on behalf of Working Party 11 on Corrosion of Steel in Concrete

38 **Corrosion of reinforcement in concrete: mechanisms, monitoring, inhibitors and rehabilitation techniques**
Edited by M. Raupach, B. Elsener, R. Polder and J. Mietz on behalf of Working Party 11 on Corrosion of Steel in Concrete

39 **The use of corrosion inhibitors in oil and gas production**
Edited by J. W. Palmer, W. Hedges and J. L. Dawson on behalf of Working Party 13 on Corrosion in Oil and Gas Production

40 **Control of corrosion in cooling waters**
Edited by J. D. Harston and F. Ropital on behalf of Working Party 15 on Corrosion in the Refinery Industry

41 **Metal dusting, carburisation and nitridation**
Edited by H. Grabke and M. Schütze on behalf of Working Party 3 on Corrosion by Hot Gases and Combustion Products

42 **Corrosion in refineries**
Edited by J. D. Harston and F. Ropital on behalf of Working Party 15 on Corrosion in the Refinery Industry

43 **The electrochemistry and characteristics of embeddable reference electrodes for concrete**
Prepared by R. Myrdal on behalf of Working Party 11 on Corrosion of Steel in Concrete

44 **The use of electrochemical scanning tunnelling microscopy (EC-STM) in corrosion analysis: reference material and procedural guidelines**
Prepared by R. Lindström, V. Maurice, L. Klein and P. Marcus on behalf of Working Party 6 on Surface Science

45 **Local probe techniques for corrosion research**
Edited by R. Oltra on behalf of Working Party 8 on Physico-Chemical Methods of Corrosion Testing

46 **Amine unit corrosion survey**
Edited by J. D. Harston and F. Ropital on behalf of Working Party 15 on Corrosion in the Refinery Industry

47 **Novel approaches to the improvement of high temperature corrosion resistance**
Edited by M. Schütze and W. Quadakkers on behalf of Working Party 3 on Corrosion by Hot Gases and Combustion Products

48 **Corrosion of metallic heritage artefacts: investigation, conservation and prediction of long term behaviour**
Edited by P. Dillmann, G. Béranger, P. Piccardo and H. Matthiesen on behalf of Working Party 4 on Nuclear Corrosion

49 **Electrochemistry in light water reactors: reference electrodes, measurement, corrosion and tribocorrosion**
Edited by R.-W. Bosch, D. Féron and J.-P. Celis on behalf of Working Party 4 on Nuclear Corrosion

50 **Corrosion behaviour and protection of copper and aluminium alloys in seawater**
Edited by D. Féron on behalf of Working Party 9 on Marine Corrosion

51 **Corrosion issues in light water reactors: stress corrosion cracking**
Edited by D. Féron and J-M. Olive on behalf of Working Party 4 on Nuclear Corrosion

52 **Progress in Corrosion – The first 50 years of the EFC**
Edited by P. McIntyre and J. Vogelsang on behalf of the EFC Science and Technology Advisory Committee

53 **Standardisation of thermal cycling exposure testing**
Edited by M. Schütze and M. Malessa on behalf of Working Party 3 on Corrosion by Hot Gases and Combustion Products

54 **Innovative pre-treatment techniques to prevent corrosion of metallic surfaces**
Edited by L. Fedrizzi, H. Terryn and A. Simões on behalf of Working Party 14 on Coatings

55 **Corrosion-under-insulation (CUI) guidelines**
Prepared by S. Winnik on behalf of Working Party 13 on Corrosion in Oil and Gas Production and Working Party 15 on Corrosion in the Refinery Industry

56 **Corrosion monitoring in nuclear systems**
Edited by S. Ritter and A. Molander

57 **Protective systems for high temperature applications: from theory to industrial implementation**
Edited by M. Schütze

58 **Self-healing properties of new surface treatments**
Edited by L. Fedrizzi, W. Fürbeth and F. Montemor

60 **Methodology of crevice corrosion testing for stainless steels in natural and treated seawaters**
Edited by U. Kivisäkk, B. Espelid and D. Féron

61 **Inter-laboratory study on electrochemical methods for the characterisation of CoCrMo biomedical alloys in simulated body fluids**
Edited by A. Igual Munoz and S. Mischler

Preface

This paper reports the results of an inter-laboratory investigation evaluating the reproducibility of electrochemical test protocols commonly used in research assessing the corrosion behaviour of biomedical CoCrMo alloys used for artificial joints.

Fifteen corrosion laboratories from seven European countries and from Japan successfully participated in this study endorsed by the COST533 Action 'Materials for Improved Wear Resistance of Total Artificial Joints' and by the European Federation of Corrosion, WP18 Biotribocorrosion.

Despite the good qualitative agreement on the general corrosion behaviour found among the participants, shortcomings in experimental protocol and data extraction procedures caused much scatter in the corrosion rates determined for the investigated alloy. From a practical point of view, this work has shown that the present state-of-the-art does not allow discrimination between negligible and large releases of hazardous corrosion products into the body.

This work stresses the importance of developing improved corrosion test protocols for reliable prediction of long-term material release from biomedical implants and to gain a deeper scientific understanding of the reactions involved.

List of symbols

Surface area (m^2)	A
Capacitance ($C \cdot V^{-1}$ or $C \cdot V^{-1} \cdot m^{-2}$)	C
Electrode potential (V)	E[1]
Breakdown potential (V)	E_b[1]
Corrosion potential (V)	E_{corr}[1]
Current (A)	I
Current density (A/cm^2)	i
Anodic current (A/cm^2)	I_{anodic}
Cathodic current (A/cm^2)	$I_{cathodic}$
Corrosion current density (A/cm^2)	i_{corr}
Passivation current density (A/cm^2)	i_p
Passive current density (A/cm^2)	i_{pp}
Faraday's constant (C/mol)	F
Atomic mass of the metal (g/mol)	M
Metal mass oxidised (g)	m
Mean value	MV
Oxidation valence	n
Open Circuit Potential (V)	OCP
Polarisation resistance ($\Omega \cdot cm^2$)	R_p
Solution resistance ($\Omega \cdot cm^2$)	R_s
Inter-laboratory variance	S_L^2
Repeatability	S_r^2
Reproducibility	S_R^2
Time (s)	t
Temperature (°C)	T
Corrosion rate ($mg \cdot dm^{-2} \cdot year^{-1}$ or $\mu m \cdot year^{-1}$)	V_{corr}
Frequency (s^{-1})	w
Impedance ($\Omega \cdot cm^2$)	Z
Anodic Tafel coefficient (mV)	β_a
Cathodic Tafel coefficient (mV)	β_c
Phase angle (°)	θ
Standard deviation	σ
Density (g/cm^2)	ρ

[1] In this paper, all potentials are given with respect to the standard hydrogen electrode.

1
Introduction and rationale

All metals and alloys are subjected to corrosion when in contact with body fluid as the body environment is very aggressive owing to the presence of chloride ions and oxygen. A variety of chemical reactions occur on the surface of a surgically implanted alloy. The metallic components of the alloy are oxidised to their ionic forms and dissolved oxygen is reduced to hydroxide ions. The rate of attack by general corrosion is very low due to the presence of passive surface films on most of the metallic implants that are currently used. Nevertheless, corrosion of implants has clinical consequences and there is a need to gain a better understanding and control of this phenomenon. Robust corrosion investigation protocols are needed. The goal of this inter-laboratory comparison is to evaluate the robustness of electrochemical practices commonly used for the study of biomaterials corrosion.

1.1 Corrosion and biomedical alloys

Corrosion is an irreversible interfacial reaction of a material with its environment, resulting in the loss of the material or in the dissolving of one of the constituents of the environment into the material. The most familiar example of corrosion is the rusting of steel due to the chemical transformation of iron into loose iron oxide by chemical reaction with water and oxygen. The corrosion of steel constitutes an enormous technological and economic issue as every second, nearly 5 tons of steel are destroyed worldwide by corrosion. A welcomed appearance of corrosion is the formation of the decorative greenish patina on copper roofs. Other corrosion reactions are less apparent but critically affect the performance of materials.

On certain metals such as titanium and stainless steel, a 1–2 nm thick compact surface oxide layer (passive film) forms by reaction with water that significantly reduces the reactivity of the underlying metal with the environment. Passive films also play a significant biological role as in the case of titanium alloys that owe their excellent biocompatibility to the properties of the titanium oxide passive film. Classical metal alloys used in implants are titanium base alloys, iron–chromium-based stainless steels and cobalt–chromium alloys. All of these alloys are passive as they spontaneously form titanium oxide (titanium alloys) or chromium oxide passive films in body fluids that provide outstanding corrosion resistance.

Although passive films reduce the corrosion rate of biomedical alloys, they do not entirely suppress it. Metal atoms can still be oxidised to metallic ions and diffuse through the passive film. Furthermore, passive films are thin and can be easily destroyed by scratching or rubbing against a solid counterpart, for example, in joint replacements. In this case, severe corrosion (fretting corrosion, tribocorrosion) takes place until the passive film eventually forms again. Other forms of localised corrosion (crevice corrosion, galvanic corrosion, pitting corrosion) have been reported in the literature [1], however, only for specific clinical cases. From a theoretical point of view, it is clear that the corrosion of metallic implants does occur. Although the limited corrosion rate of passive metals is not expected to affect the mechanical integrity of the implant, it implies a continuous release and accumulation of metallic ions in the body that can adversely affect the patient in the long term.

1.2 Corrosion of biomedical implants

'*Does corrosion matter?*' was the provocative title of an editorial in the *Journal of Bone and Joint Surgery* written by Professor J. Black (School of Medicine, University of Pennsylvania) in an attempt to appraise the clinical importance of the in-vivo metal release from implants [2]. Based on a survey of the published evidence for implant corrosion (11 papers from 1960 to 1987 on corrosion of joint replacements) and associated pathologies, Black came to the conclusion that "*Yes, corrosion does matter. All metallic implants corrode. The corrosion products are biologically active. Patients do exhibit symptoms linked to corrosion products from implants*". Following this clear conclusion, a number of research studies were devoted to the investigation of the in-vivo and in-vitro corrosion of biomedical alloys, in particular of stainless steel and titanium alloys.

Significant interest in cobalt–chromium–molybdenum alloys (CoCrMo) and their corrosion behaviour has developed in recent years. CoCrMo biomedical alloys have been used for orthopaedics and dental prostheses for more than five decades with outstanding results, with 10-year survival rates generally exceeding 90%. As a result of these excellent results, the number of implanted joints (total hip and knee prostheses) is increasing at an annual rate of 10%. As an example, the number of total knee prostheses implanted annually worldwide was estimated in 2008 to be about 1.5 million. Even though the in-vivo corrosion resistance of the alloys is exceptional, the passive corrosion of these alloys is sufficiently high to allow detection of increased concentrations (a few ppb) of Co and Cr in the blood, serum, and urine of patients having such prostheses. As the two metals are known to be allergenic, within the first 5 years, a small number (estimated around 0.1%) of patients may develop an allergic reaction due to their prostheses which can only be treated by the removal of their CoCrMo devices.

To be able to minimise these allergic reactions, knowledge of the in-vivo and, as a prerequisite, in-vitro corrosion behaviour of these CoCrMo biomedical alloys is mandatory. A better comprehension of this behaviour can only be achieved if accepted and robust corrosion protocols are available.

1.3 Rationale of the inter-laboratory study

The question about the robustness of existing corrosion protocols for in-vitro investigation of CoCrMo alloys was raised during a meeting of the COST 533 Action (Materials for Improved Wear Resistance of Total Artificial Joints) held in Vienna in January 2007. In an attempt to test the existing protocols, it was decided to launch the 'Inter-laboratory study on electrochemical methods for characterisation of CoCrMo biomedical alloys in simulated body fluids' aimed at assessing the reproducibility among different laboratories of electrochemical measurements typically used for in-vitro investigation of corrosion processes.

Secondary targets were to define improved electrochemical protocols for corrosion testing of biomedical materials and to evaluate the network capabilities of laboratories involved in the electrochemical characterisation of biomedical alloys as a prerequisite for future joint research projects.

The coordination was undertaken by Professor A. Igual Munoz (Universidad Politécnica de Valencia, Spain) and Dr S. Mischler (Ecole Polytechnique Fédérale de Lausanne EPFL, Switzerland) with the support of the COST 533 Action and the endorsement of the European Federation of Corrosion (Working Party 18). Thirteen laboratories from seven European countries and two laboratories from Japan participated in the study.

A test protocol specifying sample preparation, tests conditions and results to be extracted was distributed together with CoCrMo alloy samples. Experiments included the determination of potentiodynamic polarisation curves and electrochemical impedance spectra for a biomedical CoCrMo alloy in phosphate buffered solutions with and without albumin. Most of the laboratories carried out the experiments before the end of 2007 and preliminary results were presented at the EUROCORR 2007 conference in Freiburg im Breisgau, Germany. The final results were reported in two technical documents dated March 2008 and July 2008 and presented at the COST533 meeting held in Athens on 6–8 October 2008. At that meeting, it was decided to finalise the conclusions of the inter-laboratory study in a comprehensive final report and successively in papers to be published in scientific journals.

Accordingly, this final document is intended to present the rationale of this inter-laboratory study, to summarise the main results and to assess their impact on corrosion and biomedical practice. First, the state-of-the-art related to the corrosion of CoCrMo biomedical alloys is reviewed. Second, the protocol and statistical analysis of the results obtained are presented. The discussion section is centred on repeatability and reproducibility issues as well as on the impact of the study on clinical practice and corrosion science. Possible improvements in the protocol are discussed.

2
State-of-the-art

The objective of this chapter is to identify experimental techniques used by researchers to evaluate corrosion and to summarise the manipulation procedures and interpretation methods of electrochemical data for corrosion studies of biomedical alloys.

2.1 Literature search

The literature search has been carried out using the keywords 'Corrosion' and 'CoCrMo' in the following databases:

- ISI web of knowledge
- Electrochemical Society (ECS)
- Medline
- Science Direct.

The quantitative outcome of papers versus year is shown in Figure 2.1. It is possible to observe a significant increase in published papers from 2001. Table 2.1 presents a representative list of 26 papers [3–28] published in the last 13 years describing *in-vitro* corrosion studies used for the investigation of CoCrMo biomedical alloys in aqueous solutions.

Electrochemical techniques in combination with instrumental analysis (XPS, Raman, ICP, etc.) were used to investigate the corrosion mechanisms of CoCrMo, in particular, the passivity, transpassivity, and adsorption of organic molecules such as albumin. The investigated mechanisms are included in Table 2.1.

2.2 Experimental techniques

2.2.1 Measurement of open-circuit potential

This technique consists of recording the open-circuit potential, i.e. the potential difference spontaneously established between the working electrode (the metal being investigated) and a reference electrode placed in the solution close to the working electrode. The measurement of the open-circuit potential (OCP) during corrosion tests is a very simple technique that allows gathering of information on the surface state of the metal/alloy [14,20,26,27].

Immersion times during which the open-circuit potential was recorded by different researchers varied from several minutes (30 min [10,14], 20 min [20,26], 10 min [27], 60 min [12]) to several hours (24 h [3], 32 h [16]) or several days (140 h, ≈6 days [6,7], 2200 h, ≈92 days [19]). In some cases, the stable potential established during immersion before the electrochemical measurements (i.e. potentiodynamic, potentiostatic) was considered to be the open-circuit potential (OCP).

Open-circuit potential measurements were used by several authors [3,16] to draw a relative comparison of the nobility of the alloys in the test situation and to construct a galvanic series. Reclaru *et al.* [16] observed that this electrochemical variable is not

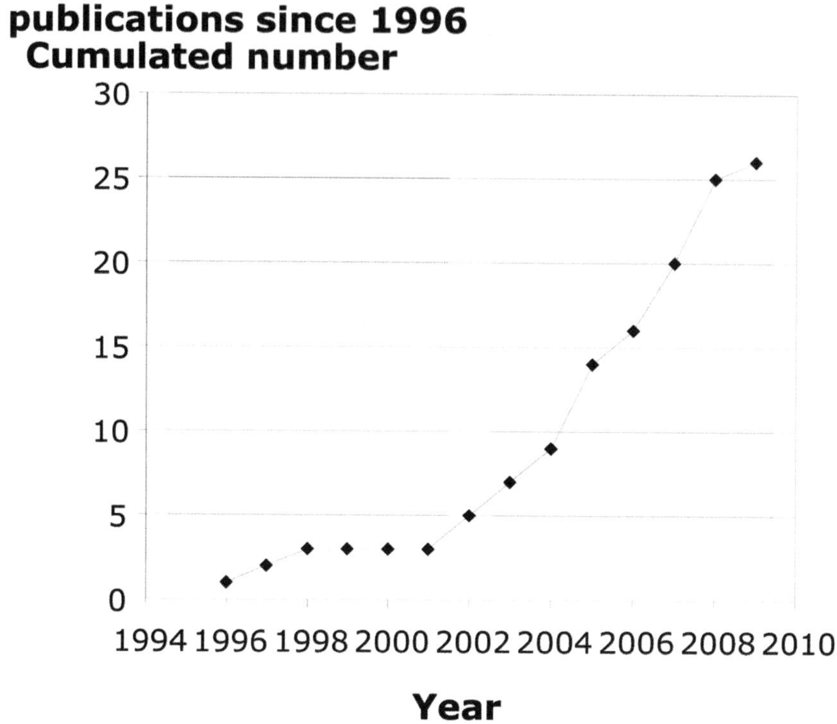

2.1 Cumulative number of papers on corrosion of CoCrMo alloys since 1996 versus year

specific to reversible phenomena and therefore Nernst's equilibrium equation is no longer valid. The nature of the metal–solution interface varies with time and consequently, the open-circuit potential is no longer characteristic of the metal. It also depends on the experimental conditions, particularly on the electrolyte composition, the temperature and the oxygen content of the electrolyte, and on the surface state of the metal.

Open-Circuit Potential (OCP) varies depending on the solution and the change is attributed to both the anodic dissolution of the implant materials and the reduction reaction (mainly the oxygen reduction reaction). Contu and colleagues [6–8,13] used OCP as an indicator of the influence of the solution chemistry on the cathodic reaction. However, additional experimental techniques (e.g. Electrochemical Impedance Spectroscopy (EIS)) are commonly used to corroborate the conclusions obtained from the OCP analysis. No clear corrosion mechanisms can be obtained from the simple measurement of the open-circuit potential. It is also worth noting that this technique does not give information on the kinetics of the corrosion reactions.

Some confusion can be found in the terminology. Sometimes the open-circuit potential is also called the corrosion potential. In this paper, we use the terminology proposed by Luthy *et al.* [3] who established that the open-circuit potential is the measured value of the potential (as described above). The corrosion potential is a value extracted from polarisation curves and corresponds to the potential where the current changes sign from cathodic (negative) to anodic (positive).

Table 2.1 Papers published since 1996 on in-vitro corrosion experiments for CoCrMo testing

Reference	Year	CoCrMo	Environment[a]	Electrochemical technique[b]	Surface analysis[c]	Investigated parameter	Corrosion mechanism
[3]	1996	HS-21 Precicast SA	Art. saliva 37°C, pH 5	I, II, IV	A	Galvanic current Galvanic potential OCP vs time	Galvanic Crevice
[4]	1997	ASTM F75	PBS PBS+10%FBS PBS+10%BSA pH 7 and pH 2 Room temperature	III		E_{corr}, i_{corr} *i* vs *t* *I* peak current	Depassivation Repassivation
[5]	1998	Remanium GM800 Dentarum ZrO_2 coating	Art. saliva	III, IV	A, B, D AAS	E_{corr}, i_{corr} metal release	Active, passive and transpassive dissolution
[6]	2002	Cast Sulzer	FBS+antibiotic Na_2SO_4 pH 7, 25°C	I, V		OCP vs *t* *C* and *R* vs *t*	Adsorption Passive dissolution
[7]	2003	Cast Sulzer Sandblasted samples	FBS+antibiotic Na_2SO_4 pH 7, 25°C	I, V		OCP vs *t* *C* and *R* vs *t*	Adsorption Passive dissolution
[8]	2003	Cast Sulzer	Buffer citrate pH 3–6 PBS, pH 7 Buffer borate pH 8–10 25°C	I, IV		OCP	Anodic dissolution
[9]	2003	ASTM F75	SPS+glucose+ tri-sodium citrate dehydrate pH 7.5	III, IV	C	Surface composition	Passive and Transpassive dissolution

Table 2.1 Continued

Reference	Year	CoCrMo	Environment[a]	Electrochemical technique[b]	Surface analysis[c]	Investigated parameter	Corrosion mechanism
[10]	2004	Co28Cr6Mo (cast Protasul-2 ASTM F75) DLC coatings	NaCl	IV	A Raman	E_{rest}, E_{brk}, i_{brk}, E_{back}, i_{back}, i_{max}	Anodic dissolution Localised
[11]	2004	Co28Cr6Mo (Protasul-20) ASTM F75	NaCl, SBF pH 7.4, 37°C	III, IV, V	C ICP–MS	i vs t OCP Surface composition	Passive dissolution
[12]	2004	CoCrMo ASTM F75 CoNiCrMo ASTM F562	SPS Addition EDTA, citrate	IV, V	i_{corr}, E_{corr}	R_p, E_{corr} Equivalent Electrical Circuit vs solution chemistry	Active, passive and transpassive dissolution
[13]	2005	Cast Sulzer	Buffer citrate pH 4 PBS pH 7 NaOH pH 14 FBS 25°C	I, IV		OCP i_{corr}	Anodic dissolution
[14]	2005	Co29Cr6Mo Forging ratios Co29Cr6Mo1Ni ASTM F75-92	NaCl Hank's E-MEM+FBS pH 5.5, 7.3, 8.3 37°C	I, IV	B, C	Crystalline structure Surface composition OCP vs t i_p	Passive and transpassive dissolution
[15]	2005	HC and LC ASTM F1537-94 ISO 5832-12	Deionised water pH 7.7; PBS pH 7.4 Synovial fluid pH 7.8 37°C		C, D, E Et-AAS		Passive dissolution
[16]	2005	Co28Cr6Mo ASTM F75	Art. bone fluid (Burks and Peck) 37°C	I, III, IV	ICP–MS	OCP, R_p, i_{corr}, E_b, i vs t	Active, passive and transpassive dissolution

Table 2.1 Continued

Reference	Year	CoCrMo	Environment[a]	Electrochemical technique[b]	Surface analysis[c]	Investigated parameter	Corrosion mechanism
[17]	2006	Co30Cr6Mo Pure metals	Hank's+glucose pH 6.8, pH 2, 25°C	IV, V	HR ICP–MS	E_{corr}, R_p, C_{dl}	Passive and active dissolution Pitting
[18]	2006	Co26Cr6Mo ISO5832-12 nitrogen ion implanted	SBF 37°C, pH 7.4		A, B, F ETAAS ICP–OES	Metal release	Passive dissolution
[19]	2007	Endocast SL (ISO 5832-12) Implanted Na-ions	SBF 37°C, pH 7	IV, V	C, G, H	R_p, E_b Morphology studies	Pitting
[20]	2007	CoCrMo Protasul-20	NaCl NaCl+albumin PBS PBS+albumin pH 7.4, 37°C	I, III, IV, V	C, D	OCP, i_{corr}, E_{corr}, R_p, C_{dl} Surface composition	Passive dissolution Adsorption
[21]	2007	Pure Co Co30Cr6Mo (Goodfellow)	Hank's+glucose	IV, V		Cathodic peaks Equivalent electrical circuit	Active and passive dissolution Pitting
[22]	2007	HC, LC CoCrMo	DMEM NaCl	I, III, IV	C	E_{corr}, i vs t	Active and passive dissolution
[23]	2009	Co30Cr6Mo (Goodfellow) Pure metals	Hank's pH 6.8	IV, V		Equivalent Electrical Circuit (EEC)	Passivity
[24]	2008	HC CoCrMoNiFe LC CoCrMoNiFe	α-medium, PBS Calf serum, Art. saliva Ringer's, NaCl Lactic acid L-cysteine HCl	IV	ICP–MS	Metal ion concentration Passive current density and potential	Anodic dissolution

Table 2.1 Continued

Reference	Year	CoCrMo	Environment[a]	Electrochemical technique[b]	Surface analysis[c]	Investigated parameter	Corrosion mechanism
[25]	2008	Wt% Co 65.21, Cr 27.29, Mo 5.54, Mn 0.65, Si 0.69, Ni 0.13, Fe 0.22, N 0.18, C 0.089	Calf serum PBS K_2CrO_4 pH 7.4	IV, V	C UV–Vis	OCP evolution E_{corr}, i_{corr} EEC	Passive dissolution Adsorption
[26]	2008	AISI 316L Co28Cr6Mo ISO 5832-12	NaCl NaCl+albumin PBS PBS+albumin pH 7.4, 37°C	I, III, IV, V		E_{corr}, i_{corr}, E_p; i_{pp}, E_{rp}, i_{rp} EEC	Active and passive dissolution Pitting
[27]	2009	Wt% Co (Bal) Cr 28.4, Mo 5.39, Mn 0.38, Si 0.8, Ni 0.22, Fe 0.22, N 74 ppm, C 0.259	BS (Sigma)	I, III, IV, V		E_{corr}, i_{corr}, E_b; i_{pp} I vs t EEC	Passive and transpassive dissolution
[28]	2009	CoCrMo ASTM F75	NaCl PBS 25% BS 50% BS pH 7.4	IV, VI	E, F, J		

[a] **Environment**: PBS, phosphate buffered solution; FBS, fetal bovine serum; BSA, bovine serum albumin; SPS, simulated physiological solution; BS, bovine serum; SBF, simulated body fluid; E-MEM, Eagle's Minimum Essential Medium; HBSS, Hank's balanced salt solution; HA, hyaluronic acid; CS, calf serum; DMEM, Dubelcco's Modified Eagle's Medium.

[b] **Electrochemical technique**: I, Corrosion Potential; II, Galvanic cells (Zero Resistance Ammetry); III, Potentiostatic; IV, Potentiodynamic; V, Electrochemical Impedance Spectroscopy; VI, Electrochemical Noise.

[c] **Surface analysis**: A, SEM (scanning electron microscopy); B, XRD (X-ray diffraction); C, XPS (X-ray photoelectron spectroscopy); D, AES (auger electron spectroscopy); E, FIB (focused ion beam); F, AFM (atomic force microscopy); G, TEM (transmission electron microscopy); H, SIMS (secondary ion mass spectrometry); I, OM (optical microscopy); J, FEG-SEM (field emission gun scanning electron microscopy).

2.2.2 Galvanic cells (Zero Resistance Ammetry)

The galvanic cell technique consists of two working electrodes placed at a certain separation and connected to a zero-resistance ammeter measuring the galvanic current. The galvanic current should ideally represent the anodic current between the less noble material and the more noble material. The current flowing in the galvanic cell depends on the potentials established on both electrodes as well as on the electrical resistance of the electrolyte, i.e. on electrolyte conductivity and the distance between both samples. As with the corrosion potential technique, this method has the advantage of simplicity and of working at the corrosion potential, i.e. under similar conditions as in engineering systems. In addition, due to its semi-quantitative character, it allows comparison of material couples. For example, using galvanic cell measurements and recording galvanic current and common potential for 24 h, Luthy *et al.* [3] found that CoCrNi alloy cannot be brazed with a gold braze in removable partial dentures (RPDs). In the case of CoCrMo/CoCrNi and CoCrMo/Co pairs, the CoCrMo acted as the cathode. However, to improve understanding of the specific kinetic parameters of the galvanic cell, predictive methods were also examined (i.e. Evan's plots, application of the mixed potential theory).

2.2.3 Potentiostatic tests

In potentiostatic tests, a selected potential E is imposed on the metal samples using a three-electrode set-up including the working electrode (the metal being investigated), the reference electrode and a counter electrode (made from inert materials such as platinum or graphite). The three electrodes are connected to a potentiostat, which is an electronic device that maintains the selected potential between the working and reference electrodes by passing an appropriate current between working and counter electrodes. The current is measured at a fixed potential as a function of the time to follow the evolution of the electrochemical kinetics of the involved reactions.

The current measured, I_{measured}, during potentiostatic tests corresponds to the sum of the anodic, $I_{\text{anodic,I}}$, and cathodic, $I_{\text{cathodic,I}}$ currents due to all of the electrochemical reactions taking place on the working electrode, as described by equation 2.1

$$I_{\text{measured}} = \sum I_{\text{anodic,I}} + \sum I_{\text{cathodic,I}} \qquad [2.1]$$

Note that, by convention, cathodic currents are negative while anodic currents are positive. The potential determines the prevailing electrochemical reactions. At the corrosion potential, the measured current is zero and the anodic and cathodic reactions occur at the same rate. At cathodic potentials, i.e. well below the corrosion potential, the dissolution rate of the metal is negligible, and the measured current is determined by the kinetics of cathodic reactions. Reciprocally, for potentials well above the corrosion potential, the rate of the cathodic reactions becomes negligible and the current is determined by the kinetics of metal oxidation.

The relationship between the loss of mass of metal and current in the anodic area is given by Faraday's Law

$$m = \frac{I.M.t}{n.F} \qquad [2.2]$$

where m is metal mass oxidised during time t, I is the anodic current, F is Faraday's constant (approximately 96 500 C/mol), n is the oxidation valence and M is the atomic mass of the metal.

According to equations 2.1 and 2.2, the conversion of current into mass of oxidised metal is possible only by knowing oxidation valence and whether the oxidation of the metal is the prevailing contribution to the measured current.

In addition to the possibility of quantifying the amount of corroded metal, potentiostatic techniques allow simulation of different corrosion conditions by applying appropriate potentials. Therefore, different applied potentials (i.e. cathodic [9], passive [7,19]) for different immersion times (1 h [19], 24 h [25]) are found in the literature. For example, Milosev and Strehblow [9] studied the effect of applied potential on the composition, thickness and structure of the oxide layer formed on CoCrMo in simulated physiological solution (SPS). Therefore, potentiostatic tests were also followed by surface analysis and they observed that electrochemical oxidation of CoCrMo alloy results in the formation of a complex layer whose composition and thickness depended on the applied potential. Similarly, Hodgson *et al.* [11] analysed the electrochemical properties of the oxide layer on CoCrMo under simulated biological conditions and they found that the passive behaviour of CoCrMo is due to the formation of an oxide film highly enriched in Cr. The composition and thickness of that passive layer also depend on the applied potential.

Igual and colleagues [20,26,27] used potentiostatic tests to analyse the effect of solution chemistry on the passive behaviour of several biomaterials. The influence of proteins and inorganic anions (i.e. albumin and phosphates, respectively) was studied as a function of the applied potential and the interaction mechanisms were investigated by EIS and surface analysis.

Analysis of the metal ions released from biomedical implants at an applied potential has been performed by several authors [5]. Potentiostatic tests were used as accelerated tests to identify the most soluble species of the alloy in the solution by applying high anodic potentials to the system. These studies were also complemented by surface analysis (i.e. XPS, AES).

In scratch tests [4], a regime of depassivation (establishment of active dissolution in damaged areas) and repassivation of the activated areas determines the overall current.

2.2.4 Potentiodynamic tests

This experiment corresponds to potentiostatic tests but instead of maintaining a well-defined potential, the latter is swept at a constant rate using a function generator to drive the potentiostat. This method allows one to observe the effect of variables such as solution composition, immersion time, etc. on the different electrochemical reactions taking place depending on the potential.

Besides the existence of a standard protocol for measuring potentiodynamic curves [ASTM G5-94(2004): *Standard Reference Test Method for Making Potentiostatic and Potentiodynamic Anodic Polarisation Measurements*], experimental conditions (i.e. scan rate or range of applied potentials) varies depending on the group of researchers. From this point of view, one can find potentiodynamic experiments carried out at scan rates of: 0.1 mV/s [3,10], 0.2 mV/s [8], 0.25 mV/s [16], 0.33 mV/s [14,19,24], 1 mV/s [9,12,25,28], 2 mV/s [20,26,27], and 5 mV/s [5,11]. With respect to the range of applied potentials, in all cases, the applied potential was varied from cathodic (–1500 mV_{SCE} [20,26,27], –1000 mV_{SCE} [5,16,24], –500 mV_{SCE} [19], –500 mV_{SCE} [3]) to anodic values (+1000 mV_{SCE} [3,5], +1200 mV_{SCE} [16], +1500 mV_{SCE} [10,20,26,27], +2000 mV_{SCE} [24], +5000 mV_{SCE} [19]), although the range also presented great variability.

Some authors referred the initial potential of the anodic scan to the previously measured corrosion potential [12,14,25]. When studying the cathodic reaction, only in some cases did the applied potential decrease (i.e. from 0 mV_{SCE}) [8,13]. Although most authors determined potentiodynamic polarisation curves during their research work, no consensus or conclusion on corrosion mechanisms have been obtained from them. Corrosion potential and corrosion current densities are usually extracted from the potentiodynamic curves but no specific interpretation of these parameters has been given by most authors, nor has any interpretation been given of the corrosion current density of passive alloys.

Most authors have used the potentiodynamic polarisation curves to identify the various electrochemical domains of a specimen in the studied solution (i.e. cathodic, active–passive transition, passive and transpassive) [9,20,26,27].

In general, this method constitutes the first approach in a corrosion study and much analysis has been performed on the results obtained from the potentiodynamic polarisation curves. Comparisons between biomaterials and/or microstructure [14,16,26,27], surface treatments [19] and coatings [10] have been carried out based on electrochemical parameters extracted from the potentiodynamic polarisation curves (i.e. breakdown potentials, passive and corrosion current densities). The effect of alloying elements on the electrochemical behaviour of the CoCrMo alloy was interpreted based on the potentiodynamic polarisation curves [11,17,21]. The influence of variables such as immersion time [11] and solution chemistry [20,24–26,28] has also been studied by qualitative analysis of the potentiodynamic curves. Superimposed potentiodynamic curves are also an available option to simulate galvanic situations between different materials. In that sense, galvanic studies using the mixed potential theory were carried out by Luthy *et al.* [3]. Cathodic potentiodynamic curves were used to elucidate the electrochemical behaviour of CoCrMo alloy in the active state by the mixed potential theory [8].

Most of the authors used potentiodynamic curves combined with other techniques to obtain a deeper insight into the corrosion phenomena of CoCrMo biomedical alloys. Besides using combinations of other electrochemical techniques and surface analysis, dissolved species were also analysed following the potentiodynamic measurements [5]. Hsu and Yen [5] detected Co and Cr ion release from the accelerated tests and a comparison between alloys (coated and uncoated) was made based on the amount of metals detected.

2.2.5 Electrochemical Impedance Spectroscopy (EIS)

The impedance method consists of measuring the response of an electrode to a sinusoidal potential modulation of small amplitude (typically 5–10 mV) at different frequencies The AC modulation is superimposed either onto an applied anodic or cathodic potential [11,20,26,27] or onto the corrosion potential [6,7,17]. It serves for the measurement of uniform corrosion rates, for the elucidation of reaction mechanisms, for the characterisation of surface films [20] and for testing coatings or surface modifications [19].

In a typical impedance experiment, a potentiostat supplies the steady-state potential to which a sinusoidal perturbation is superposed by a programmable frequency generator. The latter is often built into a two-channel transfer function analyser, thus permitting simultaneous measurement of the potential and the current. The analyser determines the real and imaginary parts of the two quantities and by division calculates the impedance, Z, of the electrochemical system. Using appropriate software, the data are transferred into a computer memory and stored for subsequent drawing of impedance diagrams. The frequency range employed in impedance measurements typically goes from several milliHertz to 100 kHz or more. At low frequency, the experiments require much time and there is the risk that the electrode surface might change. The maximum usable frequency is determined by the time response of the potentiostat and the capacitances associated with the cell and the electric circuit.

Impedance data are commonly interpreted in terms of equivalent circuits. In that sense, the investigated parameters from the impedance experiments are common electrical elements (i.e. resistance, capacitance, and inductance). Contu *et al.* [6] used the evolution of resistance and capacitance values with time to interpret the interaction of serum with the metallic samples. In that sense, they related the variation of capacitance values to the adsorption of proteins. In later work [7], they also related the effective area of sandblasted samples to the ratio between the total capacitance of sandblasted samples and the capacitance of mechanically polished samples.

Electrochemical impedance spectroscopy has been used to investigate and monitor in situ changes in the passive film/electrolyte interface on CoCrMo with time of exposure of the alloy to the electrolyte solution at open-circuit potential and under applied passive potential [11,25,26]. Analysing the Bode diagrams, one may observe that the increase in the low frequency impedance values together with a more ideal capacitive behaviour with time indicate that the passive film on CoCrMo becomes more protective. Such statements, however, always depend on the solution chemistry [20].

Metikos-Hukovic and colleagues [17,21,23] used impedance experiments to analyse the electrochemical behaviour of different CoCrMo biomedical alloys and their alloying elements (Co, Cr and Mo) in simulated body fluids. According to the impedance results, they related the passive behaviour of the alloy to the passive properties of chromium. They highlighted the fact that the EIS method was extraordinarily useful in testing the corrosion resistance of metals and alloys. A small amplitude AC signal of ±5 mV does not act destructively on the system investigated, and the application of a wide range of frequencies enables the determination of the charge and potential distributions at the metal/film and film/electrolyte interfaces and the electric and dielectric properties of the surface film.

2.2.6 Electrochemical Noise (EN)

Electrochemical Noise refers to naturally occurring fluctuations in corrosion potential and corrosion current flow. Electrochemical noise monitoring can be further subdivided into electrochemical potential noise (EPN) measurements and electrochemical current noise (ECN) measurements. The combined monitoring of potential and current is particularly useful. Fluctuations in the corrosion potential can indicate a change in the thermodynamic state of corrosion processes, as for example indicated on a Pourbaix diagram, while changes in the current noise are indicative of the corrosion kinetics. The combination of potential and current noise measurements has also been used to estimate corrosion rates; the methodology is related to measuring the well-known polarisation resistance R_p.

Electrochemical noise data can provide an indication of the type of corrosion damage that is occurring; they are widely used to distinguish between general and localised attack. The severity of localised corrosion can also be gauged by the number and shape of the noise transients. This is an important advantage over other electrochemical techniques. Further fundamental advantages include the ability to monitor corrosion in low conductivity environments (for example, thin film condensation) and the absence of 'artificial' polarisation effects. Noise measurements are made in the completely 'natural' (freely corroding) state.

Only one work by Sun *et al.* [28] has made use of EN measurements. They analysed the combined effects of three-body abrasion and corrosion in an attempt to improve understanding of the depassivation/repassivation behaviour of cast CoCrMo and the extent of subsurface deformation that occurs during the wear-corrosion process. They measured the current evolution when the sample was subjected to sliding-corrosion and abrasion-corrosion processes.

2.2.7 Comparison of the methods

According to the electrochemical techniques used by researchers in corrosion studies and published in the literature since 1996, Fig. 2.2 represents the number of papers in which one or more of the six electrochemical techniques have been used for corrosion studies in-vitro.

Measurement of potentiodynamic polarisation curves is the technique most frequently employed by researchers in the corrosion field. It is considered the first approach for understanding the corrosion system. This technique is, however, mostly used with some complementary tests, i.e. EIS, corrosion potential measurements and potentiostatic tests. Corrosion mechanisms cannot always be elucidated by potentiodynamic polarisation curves alone.

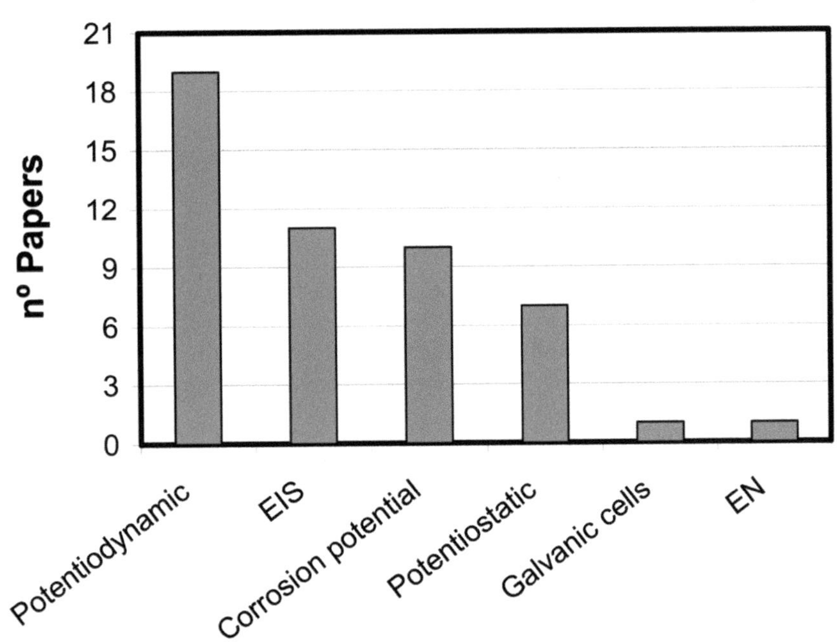

2.2 Electrochemical techniques used in papers on the in-vitro corrosion of CoCrMo alloys published since 1996

2.3 Data extraction and evaluation procedures

2.3.1 Direct Current (DC)

In DC tests, one can distinguish between measurements in which no manipulation is required (i.e. corrosion potential measurements) and measurements from which electrochemical parameters are extracted (i.e. potentiodynamic polarisation curves). There are also some techniques which do not require data extraction apart from integration of the amount of current measured (i.e. potentiostatic tests) but exceptionally, model analysis has been carried out by some authors (i.e. potentiostatic tests under scratch [4] or abrasion [28] conditions).

Therefore, data extraction from direct current measurements essentially implies the determination of parameters which give a measurement of the corrosion rate, through theoretical approaches such as the Tafel extrapolation.

In general, when authors determine i_{corr} and E_{corr} considering Tafel behaviour, only indications of the software used for the automatic extraction were found [20,26,27]. The most detailed explanation of corrosion current density determination was given by Luthy *et al.* [3]. They specified that Tafel slopes, corrosion current densities and so forth, were obtained in the region ±150 mV from the corrosion potential. They also described that the potentiostat they used (EG&G PAR model 273, Princeton Applied Research, Princeton, NJ, USA) for the electrochemical measurements included a calculating routine which used all data to perform a non-linear least-squares fit of the data to the Stearn–Geary equation. Ouerd *et al.* [25] also specified that corrosion current densities were determined by extrapolating the linear part of the cathodic curve (first plateau) to the OCP vertical axis.

Other authors, such as Kocijan *et al.* [12], evaluated I_{corr} from linear polarisation measurements using the equation

$$R_p = \frac{\beta_a \cdot \beta_c}{2.3.I_{corr} \cdot (\beta_a + \beta_c)} \qquad [2.3]$$

In the same sense, Reclaru *et al.* [16] traced polarisation curves (±150 mV vs OCP) to calculate Tafel slopes from which the corrosion current density was derived according to ASTM G59-97, with PAR Calc routine EG&G PARC Application Model 352 SoftCorr.

In general, there is a lack of consensus in the application of Tafel extrapolation.

Some other electrochemical parameters are obtained from the polarisation curves, i.e. i_{pp} and E_b. No indication was found on the criterion used for determining electrochemical parameters from potentiodynamic curves such as i_{pp} and E_b.

When analysing the galvanic current, Mixed Potential Theory is mainly employed by researchers. From this theory, galvanic current density and mixed potential were obtained from potentiodynamic curves [3].

2.3.2 Alternating Current (AC)

The theoretical interpretation of electrochemical impedance measurements must be built upon a reaction model. With the equations of the model, it is then possible to calculate the electrochemical impedance as a function of the frequency. A comparison between theoretical and experimental impedances will then lead to the confirmation or rejection of the model. In many cases, it is useful to describe the impedance of an electrochemical system in terms of an electrical equivalent circuit made of

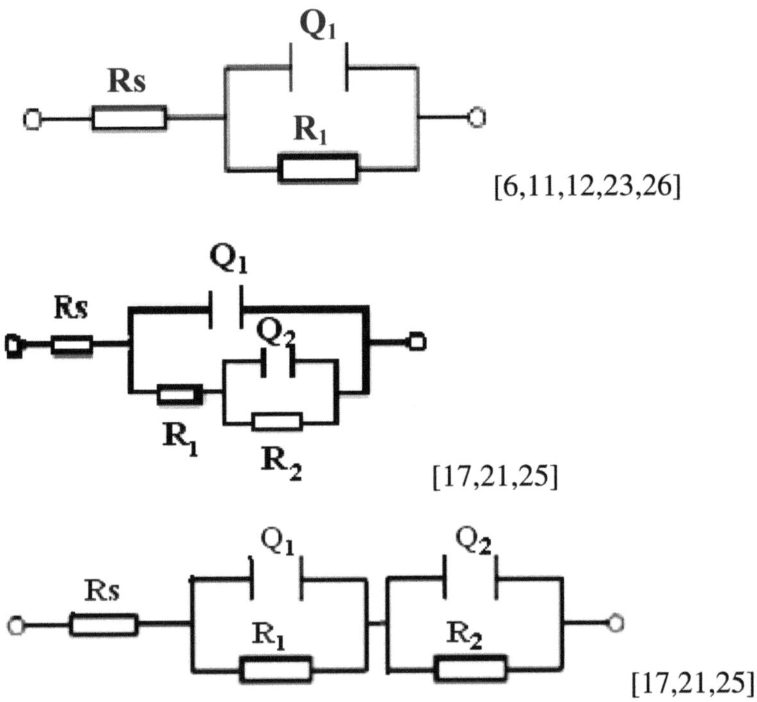

2.3 Examples of electrical equivalent circuits for describing the electrochemical interface of CoCrMo in simulated body fluids

passive elements (i.e. resistance, capacitance, inductance). From the impedance data, most researchers have used equivalent circuits to interpret experimental data [6,7,12,17,25–27].

However, there is not always consensus on the equivalent circuit to be employed for the electrochemical system. In the specific case of passive CoCrMo in simulated body fluids, several equivalent circuits have been used to describe the same system (Fig. 2.3) in which R_s is the solution resistance, R_1 is the polarisation resistance, C_1 is the interface capacitance where a constant phase element is used instead of a pure capacitor to compensate for the non-ideal capacitive response of the interface, R_2 is the resistance of the passive film and C_2 is the capacitance of the passive film.

So the interpretation of electrochemical data is not always clear. Metikos-Hukovic and colleagues [17,21,23] obtained impedance spectra for CoCrMo and its alloying elements at the OCP which were fitted to different equivalent circuits and no clear explanation of the corrosion mechanisms was obtained from them.

When fitting the impedance data to the different equivalent circuits, not all researchers specified the methodology employed. Several authors [6,12,17] stated that the spectra were fitted by a specific equivalent circuit without indicating the fitting procedure. Contu *et al.* [7] also employed more complex equivalent circuits to explain the electrochemical behaviour of a porous surface in a sandblasted CoCrMo alloy without indicating the fitting procedure. Baszkiewicz *et al.* [19] studied the R_p values obtained from the impedance data but no indication of the extraction procedure was given.

Several authors indicated no more than the software package with which the fitting was done [20,25–27], while researchers such as Metikos-Hukovic and Babic [21,23] gave a more detailed explanation when describing that the impedance spectra were analysed using complex non-linear least squares (CNLS) fitting software developed by Boukamp [29] and the ZView Version 2.8d software.

2.4 Selection criteria for the test protocol

The literature survey presented above clearly indicates that polarisation curves and EIS are the most widely used techniques for evaluating the electrochemical corrosion behaviour of CoCrMo alloys under biomedically relevant conditions. However, the test conditions varied quite widely depending on the frame and scope of the published studies. It is not surprising therefore that, depending on the test conditions, the literature sometimes reports results that cannot always be compared.

Test solutions range from pure water to complex electrolytes containing more than 40 species including adsorbable organic molecules. Clearly, the choice of electrolyte reflects the contradictory need to keep the composition as close as possible to complex body fluids while limiting the number of electrolyte components to facilitate mechanistic interpretation. The pH of the electrolytes, together with the temperature, is an important parameter controlling the conformation of large organic molecules (such as proteins) and thus their surface reactivity. Most of the electrolytes investigated are buffered solutions of constant neutral pH best approaching in-vivo conditions. In this sense, testing at 37°C should also be recommended. Another important parameter is the concentration of dissolved oxygen, as it constitutes the oxidising agent responsible for metal corrosion.

Different biomedical grade CoCrMo alloys exist. One can distinguish between low carbon (carbon content less than 0.15 wt.%) and high carbon alloys (carbon content between 0.15% and 0.3 wt.%). The main difference between the alloys is due to their microstructure, the low carbon alloys being mono phase alloys while the high carbon alloys contain chromium carbides as the second phase. Further differences are related to the process used for shaping: cast alloys are mainly used for knee joints while wrought alloys are used for components of simpler geometry (hip joints). In principle, wrought alloys exhibit a more homogeneous microstructure. The presence of second phases and an inhomogeneous microstructure are known to affect the overall electrochemical behaviour of the alloys.

Electrochemical measurements (in particular EIS) can be carried out with the tested sample under an externally imposed potential (using a potentiostat) or left at the spontaneously established potential (open-circuit potential, OCP). The former technique has the advantage that tests can be carried out under well-defined electrochemical conditions characteristic of each potential. Tests at the OCP better represent practical situations where the potential is imposed by the kinetic equilibrium between oxidation and reduction. However, one has to keep in mind that the OCP can vary with time and is dependent on very small variations in sample preparation procedure and solution composition. Not surprisingly, in many studies, both conditions were investigated in an attempt to overcome the specific limitations.

The goal of the inter-laboratory study was to evaluate the reproducibility of electrochemical corrosion techniques among different laboratories. It was therefore essential to define a common test system and to define precisely protocols for carrying out the envisaged potentiodynamic and EIS experiments. For this purpose, the use of simple experimental conditions was considered necessary and thus a low carbon, wrought alloy was selected. Furthermore, phosphate buffered solutions were chosen as simple solutions of well-defined pH although they do not closely represent the complexity of body fluids. Tests with and without albumin (as a simple, inexpensive protein) were proposed to evaluate the sensitivity of the selected corrosion techniques to biologically relevant factors. A temperature of 37°C (body temperature) was therefore adopted. The outcome of electrochemical experiments is highly sensitive to the specific protocol used. The proposed protocol was established based on the typical conditions described in the literature with regard to starting potentials and sweep rates for polarisation curve determinations and frequency ranges and timing in EIS experiments.

3 Guidelines

This chapter contains the guidelines for conducting and evaluating the experiments related to the 'Inter-laboratory study on electrochemical methods for characterisation of CoCrMo biomedical alloys in simulated body fluids'. Descriptions are given for sample and solution preparation, different experimental procedures and extraction and reporting of relevant data. Furthermore, the statistical tools used to analyse the whole set of results are described.

3.1 Introduction

Electrochemical techniques, potentiodynamic polarisation curves and electrochemical impedance spectroscopy (EIS) are established techniques in corrosion science and, as such, are of particular interest for evaluating biomaterial–environment interactions. However, the outcome of such techniques depends on a series of experimental parameters such as experimental set-up (including cell design, current distribution, ohmic resistance), tested material (composition, microstructure, surface finish), electrolyte used (including mass transport conditions and temperature), and measurement protocol (sequence of manipulations, data acquisition and extraction). There is therefore a need to validate procedures for in-vitro electrochemical experiments and to define improved experimental protocols for biomedical alloys to develop 'codes of practice' allowing for the comparison of results obtained by different laboratories.

The goal of the present inter-laboratory study was to evaluate the reproducibility of EIS and potentiodynamic measurements carried out by different laboratories using different instruments and cell configurations on a biomedical alloy in a simulated body fluid. In particular, CoCrMo implant alloys have been investigated in a phosphate buffered solution (PBS) with or without albumin (a model protein).

3.2 List of participants

The inter-laboratory study has been carried out by 15 different laboratories from Europe and Japan which are listed in Table 3.1.

3.3 Experimental conditions

3.3.1 Materials and sample preparation

The CoCrMo alloy (ASTM F1537) supplied by Surgival (Spain) has the nominal composition (wt.%): 0.037 C; 64.81 Co; 27.82 Cr; 5.82 Mo; 0.36 Si; 0.78 Mn; 0.02 Al ($S<0.0004$, $Ti <0.01$, $B<0.003$, $Zr<0.01$).

The microstructure of the CoCrMo alloy was revealed by etching the samples with Beraha III reagent (base composition: 50 g of NH_4HF_2, 600 ml of distilled water and 400 ml of concentrated HCl; with reagent composition of 1 g of $K_2S_2O_5$ for each

Table 3.1 List of participating laboratories

Institution	Country	Reference name
Centro Nacional de Investigaciones Metalúrgicas	Spain	CENIM/Lab1
Ecole Nationale Supérieure des Mines de Saint-Etienne	France	CIS/Lab2
Ecole Centrale Paris	France	ECP/Lab3
EMPA Duebendorf	Switzerland	EMPA/Lab4
Ecole Polytechnique Fédérale Lausanne	Switzerland	EPFL/Lab5
Falex Tribology N.V.	Belgium	FAL/Lab6
INSA Lyon Institut National des Sciences Appliquées de Lyon	France	INSA/Lab7
Jožef Stefan Institute	Slovenia	JSI/Lab8
Katholieke Universiteit Leuven	Belgium	KUL/Lab9
National Institute for Materials Science	Japan	NIMS/Lab10
Fundación TEKNIKER	Spain	TEK/Lab11
Tokyo Medical and Dental University	Japan	TMDU/Lab12
University of Leeds	UK	UL/Lab13
Universidade do Minho	Portugal	UM/Lab14
Universidad Politécnica de Valencia	Spain	UPV/Lab15

100 ml of base solution). The sample was immersed in the Beraha III reagent for 30 s, cleaned and rinsed with distilled water, immersed in ethanol and finally dried with compressed air. After etching, the sample was analysed by optical microscopy.

The microstructure of the sample is shown in Fig. 3.1a,b. In all cases, a dendritic structure was revealed. The same microstructures were obtained by several authors [14,30–32].

Carbides were also revealed with another reagent (Fig. 3.1c,d) [30] (base composition: 100 ml of distilled water, 4 g of $KMnO_4$ and 4 g of NaOH). The samples were immersed in the solution for 5 s, rinsed with distilled water, dried and observed directly with an optical microscope.

Samples were supplied to the participants in the machined state. Before polishing, they could be further machined to adapt their dimensions to an existing electrochemical cell.

The samples were polished to a mirror finish using 1 μm alumina paste. After polishing, they were cleaned in a water bath for 3 min, rinsed with distilled water and dried using dry, cold air. Alternatively, the samples could be polished using diamond pastes, in which case the samples were cleaned with ethanol instead of water. After cleaning, the polished samples were stored in a desiccator before the experiment and then were placed in the measurement cell which was subsequently filled with electrolyte as soon as possible.

3.3.2 Environment

Two test solutions were used as simulated body fluids. A phosphate buffered solution (PBS) containing 0.14 M NaCl, 1 mM KH_2PO_4, 3 mM KCl, 10 mM Na_2HPO_4, and a phosphate buffered solution with albumin (PBS+albumin) containing 0.14 M NaCl, 1 mM KH_2PO_4, 3 mM KCl, 10 mM Na_2HPO_4 with the addition of 1 g/l of Albumin Fraction V from bovine serum. The albumin-containing solution was prepared fresh. The pH values of the solutions were measured before the experiments to check the pH

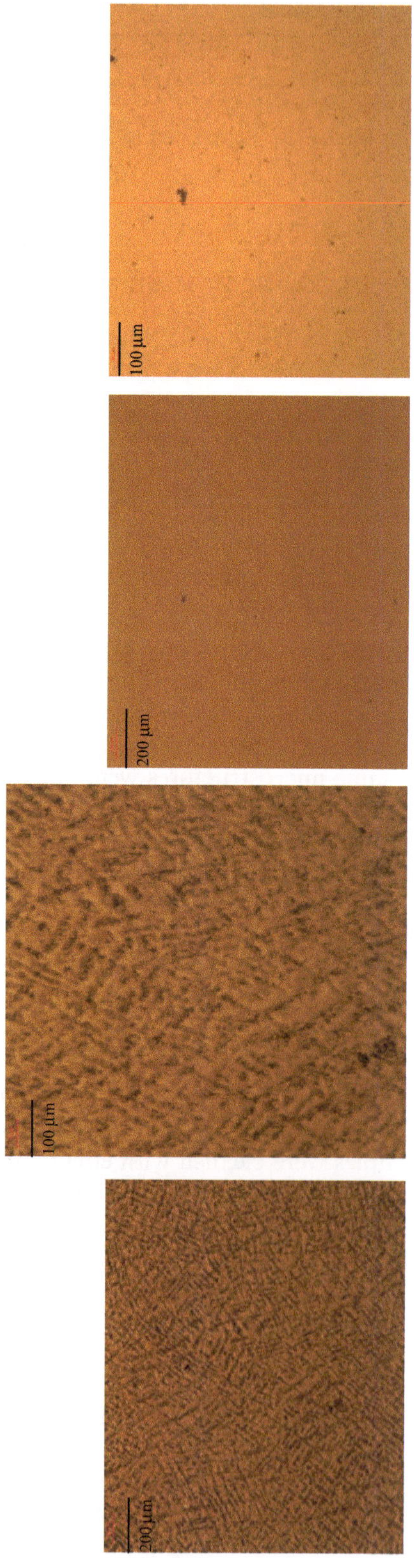

(a) (b) (c) (d)

3.1 Optical images of the microstructure of the CoCrMo alloy (a, 50× and b, 100×) and carbide precipitation (c, 50× and d, 100×)

value was 7.4. Tests were carried out at 37°C and the solutions were pre-heated to this temperature before each experiment; electrochemical measurements started when the temperature had stabilised at 37°C.

3.3.3 Electrochemical measurements

Electrochemical experiments were undertaken using a common experimental protocol. The choice of potentiostat, reference electrode, and geometry of the electrode set-up, as well as the nature of the auxiliary electrode to be used were left to the discretion of each laboratory. Each experiment involved an in-situ surface cleaning by cathodic polarisation at an imposed potential of –1.25 V_{SHE} for 5 min followed by a period at the open circuit potential.

3.4 Experiments

Three experiments were proposed. The experimental steps are illustrated in Fig. 3.2. Each experiment was repeated three times to check reproducibility.

The first experiment was a standard electrochemical test and consisted of the measurement of Potentiodynamic polarisation curves. For this, the applied potential was increased from –1.25 V_{SHE} at a constant rate (2 mV/s) to 1.5 V_{SHE} and the current needed to produce the potential was recorded in parallel. The result is a plot of the current density (i.e. the measured current divided by the electrode surface area) versus the applied potential. Some characteristic values were extracted from the polarisation curves: E_{corr} (corrosion potential), i_{corr} (corrosion current density), i_{pp} (passive current density) and E_b (breakdown potential). The characteristic values reported are illustrated in Fig. 3.3.

The second and third experiments consisted of impedance measurements under potentiostatic conditions (i.e. at an applied potential of 0.15 V_{SHE} within the passive

Experiment 1 ***Potentiodynamic curve***	**Experiment 2** ***EIS at imposed potential***	**Experiment 3** ***EIS at OCP***
Stabilisation at OCP 10 minutes	Cathodic polarisation -1.25 V vs SHE, 5 minutes	Cathodic polarisation -1.25 V vs SHE, 5 minutes
Cathodic polarisation -1.25 V SHE, 5 minutes	Stabilisation at OCP 10 minutes	Stabilisation at OCP 60 minutes
↓	Apply passive potential (0.15 V SHE)	↓
↓		↓
Potential scan	EIS at 0.15 V SHE	EIS at OCP
↓	↓	↓
Plot I vs E (polarisation curve)	Plot Nyquist diagram	Plot Nyquist diagram

3.2 Recommended procedure for surface preparation and electrochemical tests

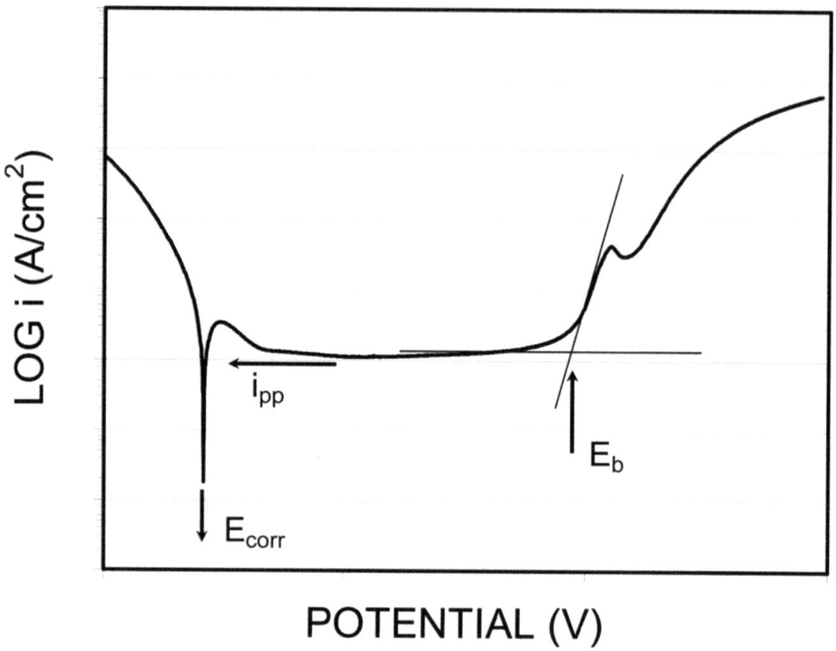

3.3 Experiment 1: Potentiodynamic polarisation curve. The characteristic parameters to be extracted are included

range and at the open circuit potential (OCP), respectively). Frequency was varied from 10^6 Hz to 10 mHz or lower frequencies, at 10 data cycles/decade with an ac amplitude of ±10 mV and the impedance response and phase angle at each frequency were recorded simultaneously. For both experiments, R_s (solution resistance) and R_p (polarisation resistance) were extracted according to Fig. 3.4.

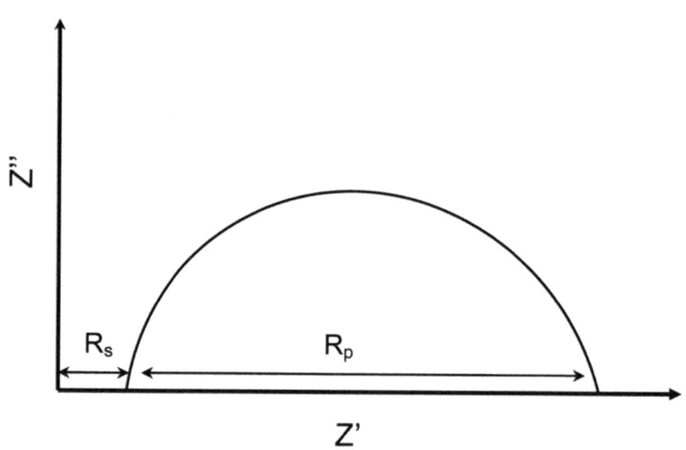

3.4 Experiment 2: Nyquist diagram. The characteristic parameters to be extracted are included

3.5 Statistical analysis

3.5.1 Introduction

Statistical analysis of the results, namely the detection and elimination of outliers, the evaluation of precision parameters (repeatability and reproducibility) and means was based on the standard ISO 5725:1994 (ISO 5725:1994, Accuracy (trueness and precision) of measured methods and results, International Organization for Standardization, Geneva, Switzerland, 1994). This recommends the participation of 8 to 15 laboratories. If fewer than eight laboratories participate in an inter-laboratory study, the reproducibility and repeatability are estimated with fewer degrees of freedom, which has consequences for the confidence intervals and uncertainties.

The following features depict the philosophy of the inter-laboratory tests:

- Interlaboratory tests produce data on the repeatability and reproducibility of the measurement results. These data represent valuable information that can be used to determine the uncertainty of a measuring procedure. The uncertainty of a measurement is mostly unknown for new test methods.
- Older standardised methods will be replaced as new analytical methods evolve. Inter-laboratory tests make a contribution to comparing the validity of traditional and newly developed test methods.
- Inter-laboratory tests should also contribute to ensuring that measurements are mutually accepted. The data are generated under routine conditions in industrial and institutional laboratories to create real reproducibility values. Day-to-day experience shows that two laboratories produce different results despite using the same method. In such cases, data from inter-laboratory tests can help in interpreting and assessing the results and changing practice when required.
- Inter-laboratory data often lead to dramatic improvements as the processing parameters for measuring methods are adjusted or even changed.
- The participants have an opportunity to test the reliability of their results and therefore also test the professionalism of their laboratory (proficiency testing).

The present inter-laboratory tests were statistically analysed using the recommendations of ISO 5275.

3.5.2 Assessment of consistency and outliers

This procedure was performed to determine and deal with inconsistent values, outliers or other irregularities. Global means and respective confidence intervals (at 95%) were calculated for each level and parameter for all of the laboratories before scrutinising the results for outliers, performed by numerical tests. First, an upper-tail Cochran test was repeatedly performed, for comparison of the inter-laboratory variances, and with consequent elimination if laboratories had inter-laboratory variability significantly above the mean (within-laboratory consistency). Grubbs tests for single and double outliers were then applied to identify laboratories with means significantly different from the others, and to eliminate them (between-laboratory variability). Care was taken to limit the number of laboratories ruled out up to 2/9 of the total, meaning up to four in this study, for each of the three samples analysed.

The Cochran test is unilateral and only the laboratories with variability above the mean are ruled out, and not the ones below. The Grubbs tests allow the detection of anomalous mean values and are performed using the values retained from the within-laboratory consistency evaluation. In the case of the single outlier test, the occurrence of one outlying result, either the smallest or the largest, is evaluated.

3.5.3 Evaluation of precision and general mean

The precision of an experimental method expresses the closeness of agreement between a series of measurements obtained from multiple sampling of the same, homogeneous sample. The precision may be considered at three levels: repeatability, intermediate precision and reproducibility. The ***repeatability*** expresses precision measured under as identical conditions as possible, i.e. over a short time, by the same analyst and with the same equipment. It is estimated normally within one laboratory, but usually can also be derived from an inter-laboratory set-up. ***Intermediate precision*** expresses within-laboratory variations: different days, different analysts, different equipment and/or different calibration. Finally, the ***reproducibility*** expresses its precision under the most diverse circumstances, including different laboratories, and thus has to be investigated in an inter-laboratory study.

In this study, after the elimination of the outliers for each parameter, values for the general mean, *m*, standard deviation σ, and variances S_L^2 (interlaboratory), S_r^2 (repeatability) and S_R^2 (reproducibility) were calculated according to standard ISO 5725: 1994.

4
Results

A summary of the results supplied by the participants is presented in this chapter. The results are statistically analysed according to the ISO5725 standard. Full results obtained by the individual laboratories are given in Annexes 1 (potentiodynamic polarisation curves) and 2 (EIS measurements).

4.1 Experimental arrangement and general comments

All laboratories carried out the tests in accordance with the general 'Experimental procedure' provided by the coordinators. Only experimental parameters linked to the specific conditions of each laboratory (potentiostats, cell geometry) were left free in the protocol. Table 4.1 shows some variables in the experimental arrangement that were not specified in the experimental procedure.

In this section, relevant comments reported by participants when carrying out and analysing the experimental data are summarised and structured according to the nature of the problems they reported:

- ***Temperature control***: Laboratories 6 and 9 could not reach the desired value of 37°C and they carried out the experiments at 23°C while Laboratory 11 could not hold the temperature at a constant value and experiments were carried out at room temperature (between 25 and 30°C).
- ***Hydrogen generation***: Laboratories 6 and 15 detected the presence of hydrogen bubbles on the working electrode during the cathodic cleaning before the electrochemical measurements which could not escape from the surface.
- ***Extraction of parameters from Experiment 1***: Several laboratories reported problems when extracting electrochemical parameters from polarisation curves. Laboratory 12 found difficulties in finding i_{pp} values and E_b values in PBS+Albumin solution because they did not find clear peaks. They also did not report i_{corr} because they were not able to extrapolate a Tafel behaviour. Laboratory 3 reported the same problem; they did not find evidence of Tafel lines on the Log(I)–E graphs. Both asked for the procedure used by the rest of the participants.
- ***Electrical noise***: Laboratory 10 found electrical noise when carrying out impedance tests at the lowest frequencies while Laboratory 8 found problems at the highest frequencies and $6.5 \cdot 10^4$ Hz was the highest frequency studied (instead of the recommended 10^6 Hz).
- ***Curve fitting in Experiments 2 and 3***: Laboratory 5 stated that "as the system under investigation does not show an ideal R–C behaviour related to the fact that the impedance was still increasing in the low frequency domain, it is impossible to determine the R_p values accurately. This can be related to the fact that the Co–Cr–Mo surface is still passivating/stabilising during the measurements. Maybe also some diffusion processes occur". Therefore, they used the following procedure to determine R_p: "for all curves, they have chosen three points to calculate the circle to estimate R_p: 4.04 Hz, 94.7 mHz and 22.88 mHz. (The calculated value depends on the choice of the points.)"

Table 4.1 Experimental arrangement of the laboratories

	Potentiostat	Area WE (cm²)	Volume (ml)	Distance RE-WE (mm)	T (°C), pH
CENIM/Lab1	Gamry	0.78	30	3	37, 7.4
CIS/Lab2	PAR 2263	1.76	20±1	8±2	37, 7.6±0.1
ECP/Lab3	Solartron 1287	1	300	10–15	7.45–7.65
EMPA/Lab4	Autolab PGSTAT30	0.785	300	3	37, 7.5
EPFL/Lab5	Autolab PGSTAT30	1.77	350	4	37
FAL/Lab6	Solartron 1287	1.38	1.5	1	23, 7.43–7.48
INSA/Lab7	PAR 2273	1.76	50	10	37, 7.6
JSI/Lab8	Autolab PGSTAT12	0.785	250	3	37, 7.57
KUL/Lab9	Solartron 1287	1	20	1	23, 7.43–7.48
NIMS/Lab10	Solartron 1287	1.13	500	5	37, 7.4–7.7
TEK/Lab11	Autolab	1	50	5	7.4
TMDU/Lab12	ALS Model 680C	0.35	250	2–5	37, 7.44–7.49
UL/Lab13	ACM Gill 12 potentiostat	0.72	80	3	37, 7.4
UM/Lab14	Radiometer PGP201, PGZ100	0.5	180	15	37, 7.4
UPV/Lab15	Solartron 1287	0.28	50	10	37, 7.4

WE, working electrode; RE, reference electrode.

Concerning impedance plots, Laboratory 3 observed "that most of them were not semi-circles as presented in the 'Experimental procedure' document, and for many of them, it was not even possible to extrapolate the diagram in the low frequency range, down to 'zero frequency' to find the real low frequency limit of the impedance. In addition, the shapes of the diagrams indicate that most of them seem to result from several relaxation processes (with several different time constants), and not from a single relaxation process (giving a single semicircle) as assumed in the 'Experimental procedure'. For a few diagrams, the semicircle can be considered as a very rough approximation, but for many others, this approximation is impossible. That means that the R_p value cannot be derived from the diagram as suggested in the 'result presentation' paragraph of the document. If the impedance diagrams have a 'complicated' shape resulting from several relaxation processes, an equivalent electrical circuit model must be used to derive transfer resistance or polarisation resistance values. They asked for more information related to the equivalent circuit to be used.

Analogously, Laboratory 12 reported that "the shape of the semicircle from the Cole–Cole plot was a bit collapsed. Curve fitting using a pure capacitor model caused large errors. The Constant Phase Element (CPE) model showed relatively good correspondence".

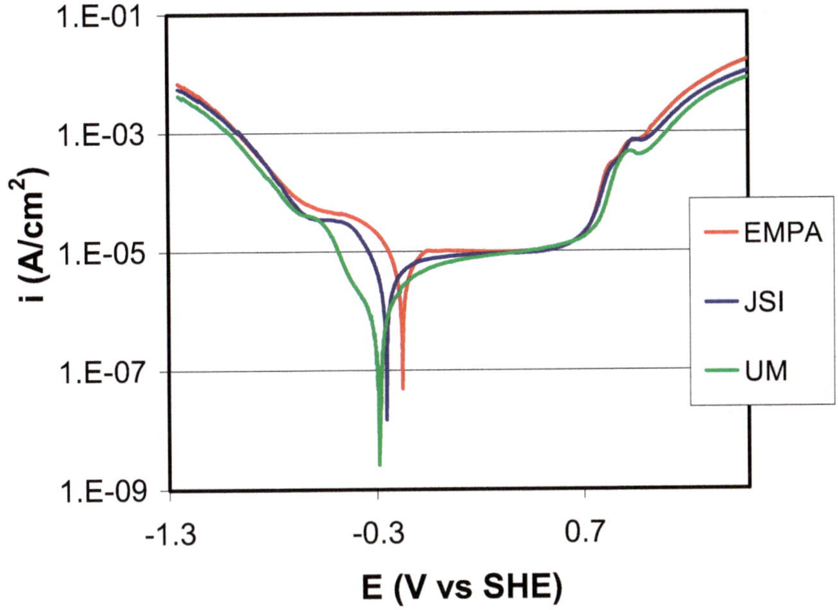

4.1 Polarisation curves measured in PBS solution by three laboratories

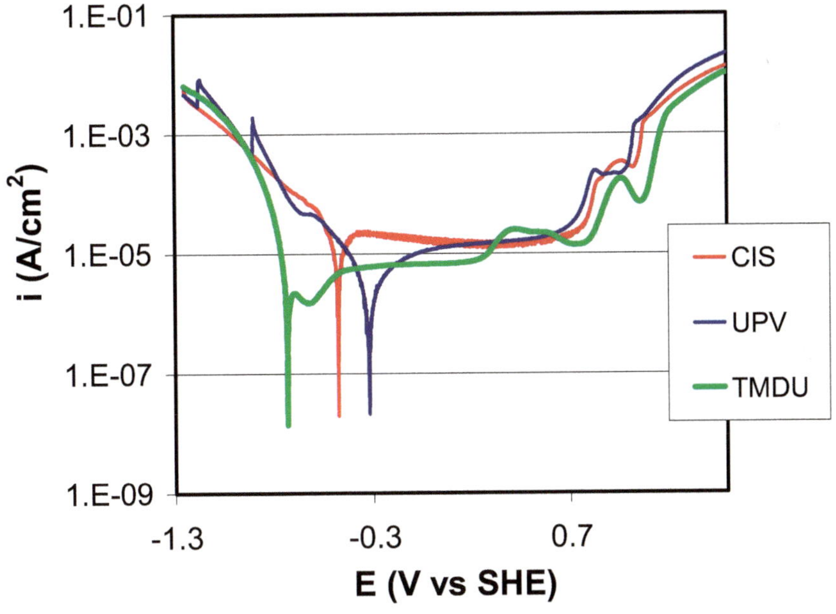

4.2 Polarisation curves measured in PBS+Albumin solution by three laboratories

Laboratories 5 and 7 made the same observation: "Impedance spectroscopy should be considered with attention in high frequencies; I think that we have an artifact for a few points relative to the capacitive contribution of our reference electrode".

4.2 Experiment 1

Figure 4.1 shows three polarisation curves measured in the absence of Albumin according to the recommended procedure. Analogously, Fig. 4.2 shows three polarisation curves for PBS+Albumin solution. Potentiodynamic curves in all cases can be divided into four different domains. The cathodic domain includes potentials from –1.3 V where the current is determined by the reduction of water and, partially, of dissolved oxygen. The potential domain between –0.7 and –0.2 V is characterised by the transition from cathodic to anodic current at the corrosion potential followed by some fluctuations of the anodic current. The third domain corresponds to the passive plateau and ranges from –0.2 up to 0.7 V. Finally, the transpassive domain is characterised by the increase in current due to transpassive dissolution of the chromium oxide as well as water oxidation. In most cases, phosphate solutions show a shoulder at approximately 0.8 V and can possibly be attributed to the formation of phosphate–chromium ion complexes that activate transpassive dissolution.

In some cases, current oscillations were observed in the cathodic domain. These oscillations are probably due to the appearance of hydrogen bubbles during the cathodic pre-treatment as reported by Laboratories 6 and 15.

Parameters extracted by the participants from the polarisation curves in PBS and PBS+Albumin are listed in Table 4.2. In the last two lines, statistical analysis carried out by the coordinators has been included. The 'MV ± σ' corresponds to the mean value and the standard deviation of each parameter without excluding any laboratory; while the 'MV ± σ' according to the standard ISO 5725:1994 corresponds to the mean value and standard deviation among all laboratories excluding some outliers (results which are considered aberrant according to the standard). The accepted number of laboratories ('Accepted labs', last line) corresponds to the number of laboratories considered for the calculation of the mean values and the standard deviation according to ISO 5725:1994.

The OCP values were measured after an immersion period of 20 min and before starting the polarisation measurements. In the PBS solution, we may observe two different groups according to the OCP values. The first one obtained OCP values around –0.3 V while the second group obtained OCP values shifted by 100 mV from the first one, around –0.2 V. In the PBS+Albumin solution, all values are around –0.3 V except one aberrant value; this value is considered to be one of the outliers according to ISO 5725:1994.

OCP values are the first parameters to be measured after the surface finishing, and this surface finishing is a critical procedure in the electrochemical tests of solid electrodes, as discussed later in Chapter 5. In fact, one of the laboratories has repeated its tests using two different lubricants for diamond paste polishing and found a shift of 200 mV in the OCP as measured according to the proposed procedure. They also observed a drift in OCP towards the same value for both lubricants. More details relating to surface finishing should be considered in an improved protocol. Alternatively, longer stabilisation times could be considered before measuring the OCP.

Corrosion currents were extracted from polarisation curves using Tafel extrapolation. Criteria for Tafel slope determination were not specified in the protocol so that the scatter found in corrosion current density probably reflects different approaches, depending on operators, in selecting the extrapolation domains. According to the ISO 5725:1994 standard, the highest number of outliers was found for i_{corr} determination.

Table 4.2 Statistical evaluation of measurements in PBS and PBS+Albumin solutions

	PBS					PBS+Albumin				
	OCP (mV)	i_{corr} (µA/cm²)	E_{corr} (mV)	i_{pp} (µA/cm²)	E_b (mV)	OCP (mV)	i_{corr} (µA/cm²)	E_{corr} (mV)	i_{pp} (µA/cm²)	E_b (mV)
Lab1	–204±8	0.7±0.2	–321±24	4.7±0.9	742±3	–200±9	1.0±0.4	–420±89	4.7±0.3	730±1
Lab2	–330±20	0.03±0.02	–270±30	7.9±0.2	760±30	300±20	0.03±0.02	–320±150	13±3	760±10
Lab3										
Lab4	–258±10	4.4±0.2	–180±5	9.9±0.8	710±10	–309±12	4.4±1.2	–427±25	9.1±0.7	690±10
Lab5	–543±100	4±1	–548±104	16.7±0.9	700±7	–490±52	1.3±0.6	–580±6	20.4±6	590±10
Lab6	–159±26	4.5±4.1	535±81	64.3±5.3	891±120	–235±52	3.9±1.5	–464±52	58.4±4.5	800±29
Lab7	–221±20		–581±139	8.7±0.2	200±29	–352±5		–469±48	6.4±0.6	1183±42
Lab8	–195±30	2.3±0.5	–255±6	7.9±0.3	701±10	–292±60	1.5±0.5	–437±48	8.3±0.4	698±5
Lab9	–196±11	2.9±2.4	–748±33	10.5±0.9	788±37	–202±9	6.3±1.2	–770±0	11.6±0.9	732±21
Lab10	–319±8	6.3±0.2	–434±35	6.4±0.3	683±9	–310±7	6.4±0.2	–434±4	5.7±0.5	675±9
Lab11	–210±87	35±19	–788±40	15±7.2	821±106	–118±56	38.5±40	–795±28	24±5	821±78
Lab12	–316±4		–557±85			–321±1		–747±16		
Lab13	–335±13	1.7±0.4	–348±29	4.5±0.2	918±12					
Lab14	–248±8	0.27±0.02	–298±30	6±0.2	737±13	–265±53	0.5±0.3	–454±34	6.2±0.7	720±13
Lab15	–45±19	1.6±0.4	–288±33	14.2±1.6	727±10	–97±22	2.6±0.5	–314±24	16.6±3.5	694±4
MV ± σ	**–256±114**	**5.5±9.5**	**–363±319**	**13.6±15.7**	**745±41**	**–220±119**	**6±11**	**–510±162**	**9.6±3.1**	**758±146**
MV ± σ ISO 5725	**–253±63**	**2.6±2**	**–363±319**	**7.4±2.1**	**728±33**	**–266±106**	**3.1±2.3**	**–535±163**	**8.1±2.9**	**710±55**
Accepted labs	**11**	**8**	**14**	**9**	**9**	**12**	**9**	**11**	**8**	**10**

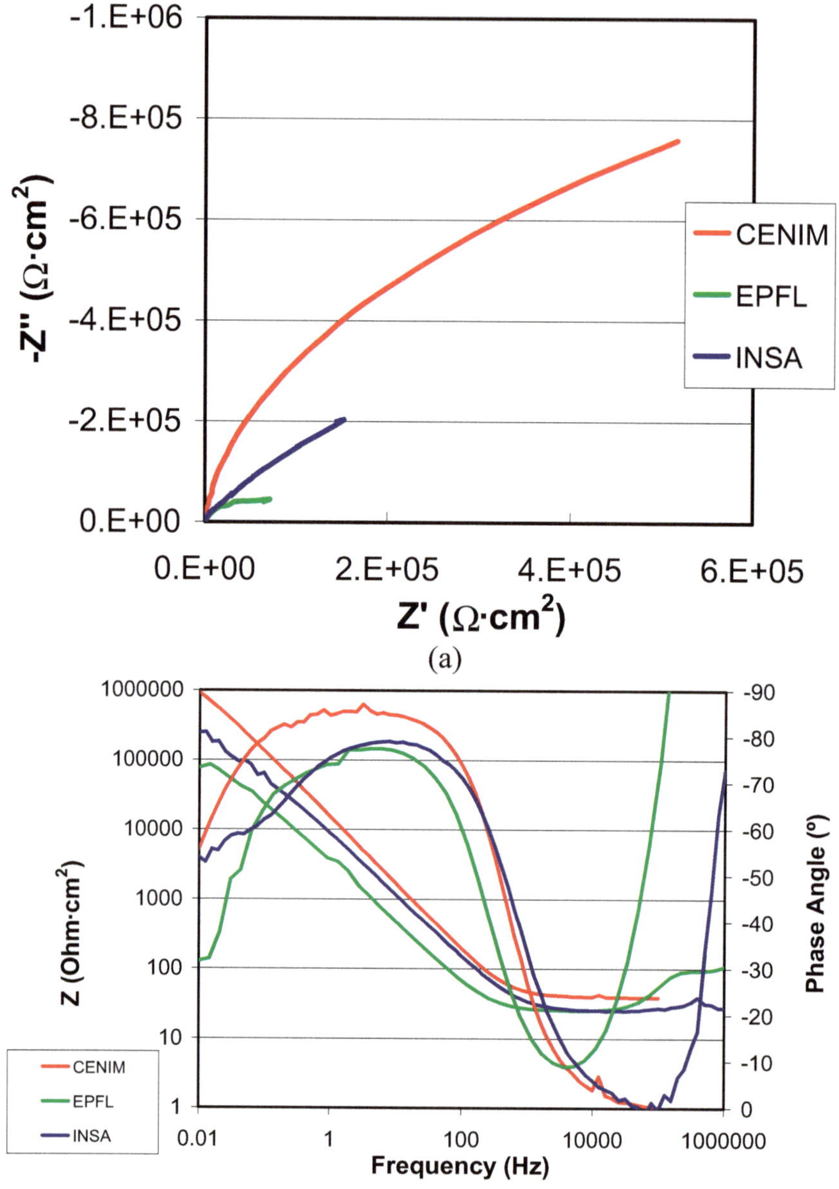

4.3 Nyquist (a) and Bode (b) plots obtained in PBS solution at 0.15 V applied potential by three different laboratories

The corrosion potential was extracted from the polarisation curves using Tafel extrapolation. This value corresponds to the potential at which the transition from cathodic to anodic current occurs. This transition potential scatters over a wide potential range. Two main groups can be distinguished in the PBS+Albumin solution: those laboratories which obtained values around –0.7 V and those which obtained values around –0.4 V. Due to the low current densities and high overvoltage governing the electrochemical behaviour of passive alloys, small variations in partial

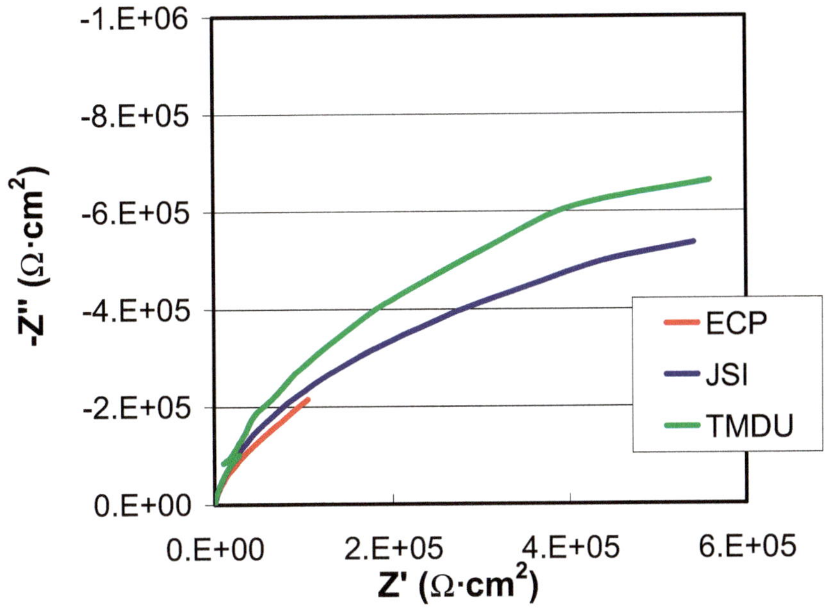

(a)

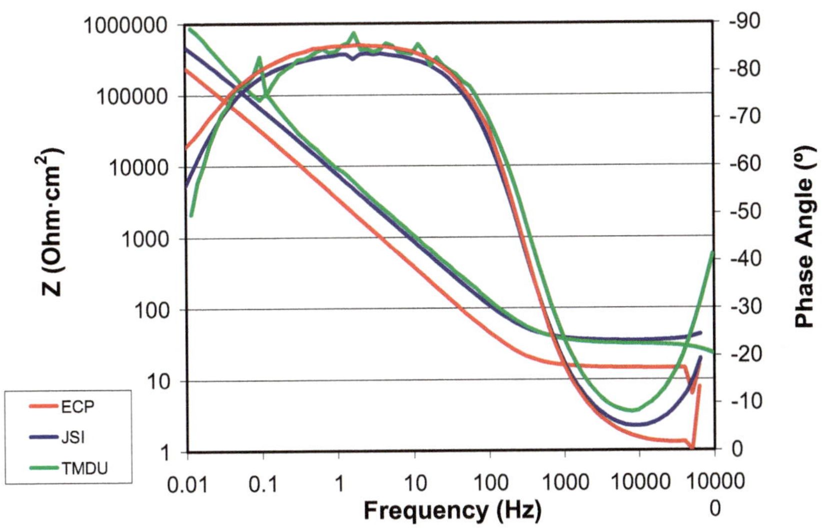

(b)

4.4 Nyquist (a) and Bode (b) plots obtained in PBS+Albumin solution at 0.15 V applied potential by three different laboratories

currents, for example, cathodic current due to oxygen reduction, may have large effects on the corrosion potential.

The passive current density i_{pp} was defined in the protocol as the lowest current density found within the passive domain. After eliminating the outliers from the statistical analysis, the scatter for i_{pp} is approximately 20%, a much lower value compared to the scatter of 100% found for i_{corr}. However, in the potentiodynamic curves (measured values) there are no significant differences in the passive domain, therefore, only the manipulation introduced the scatter.

Repeatability of E_b within the individual laboratories and reproducibility among them are very good, in fact, compared to the scatter of the other potential parameters (OCP and E_{corr}), E_b presents the lowest standard deviations. The standard deviation of the breakdown potential measurements is smaller than the deviations of the corrosion potential measurements.

4.3 Experiment 2

The aim of this experiment was to evaluate the sensitivity of the impedance measurements and the applicability to the interface characterisation in biological media.

Figures 4.3 and 4.4 show the Nyquist diagrams and Bode plots obtained at 0.15 V vs standard hydrogen electrode (SHE) in the PBS and PBS+Albumin solutions of three different laboratories, respectively.

The depressed circles observed in the Nyquist diagram indicate imperfect capacitive behaviour usually described in terms of constant phase element (CPE). The impedance spectra from approximately 0.01 Hz up to 1 kHz correspond to the electrochemical response of the interface. The solution resistance controls the spectrum above 1 kHz. However, the increase in phase angle observed at high frequency after

Table 4.3 Statistical evaluation of results obtained by each laboratory for the PBS and PBS+Albumin solutions at 0.15 V_{SHE}

	PBS			PBS+Albumin		
	i_p (μA/cm²)	R_s (Ω cm²)	R_p (kΩ cm²)	i_p (μA/cm²)	R_s (Ω cm²)	R_p (kΩ cm²)
Lab1	4.7±0.9	56±15	133±8	0.07±0.02	49±6	135±6
Lab2	3±6	31±10	1200±40	–4±7	29±3	900±20
Lab3						
Lab4	0.08±0.009	26±6	510±73	0.13±0.05	25±3	689±0.8
Lab5	7.8±2.5	23±2	108±49	0.6±0.4	25±0.6	402±38
Lab6	–67±24	–5±1	1640±2830	0.2±0.4	–4±0.4	28±20
Lab7	0.23±0.05	26±1	1030±163	0.2±0.04	26±3	822±368
Lab8	0.16±0.01	16±6	625±47	0.4±0.04	35±10	798±53
Lab9	0.18±0.19	–14±16	41±27	0.5±0.09	–26±23	44±38
Lab10	0.45±0.02	37±6	750±120	0.5±0.04	35±3	690±7
Lab11	0.2±0.04	24±3	1134±213	0.6±0.06	24±5	655±66
Lab12	0.22±0.008			0.3±0.03		
Lab13	0.85±0.3	23±2.5	2530±609			
Lab14	–18±15	41±0.6	724±110	–34±5	42±1	700±70
Lab15	0.8±0.5	56±2	989±318	1±0.2	56±2	551±443
MV ± σ	**–5.5±1.9**	**26±20**	**877±682**	**–2.6±9.5**	**27±23**	**492±333**
MV ± σ ISO 5725:1994	**0.2±0.1**	**30±11**	**604±399**	**0.4±0.2**	**35±12**	**517±280**
Accepted labs	**6**	**10**	**10**	**11**	**10**	**8**

the phase minima indicate that the impedance of the reference electrode interferes with the measurement. All participants found CPE behaviour although the extent of circle depression varied significantly. The noise at high frequencies, as observed by some participants, is likely to be introduced by the electronic circuit and the cables.

Parameters extracted by the participants from the EIS diagrams in PBS and PBS+Albumin are listed in Table 4.3.

Passivation current density is a measured value, therefore, no manipulation by participants is required. Three different groups can be distinguished according to the

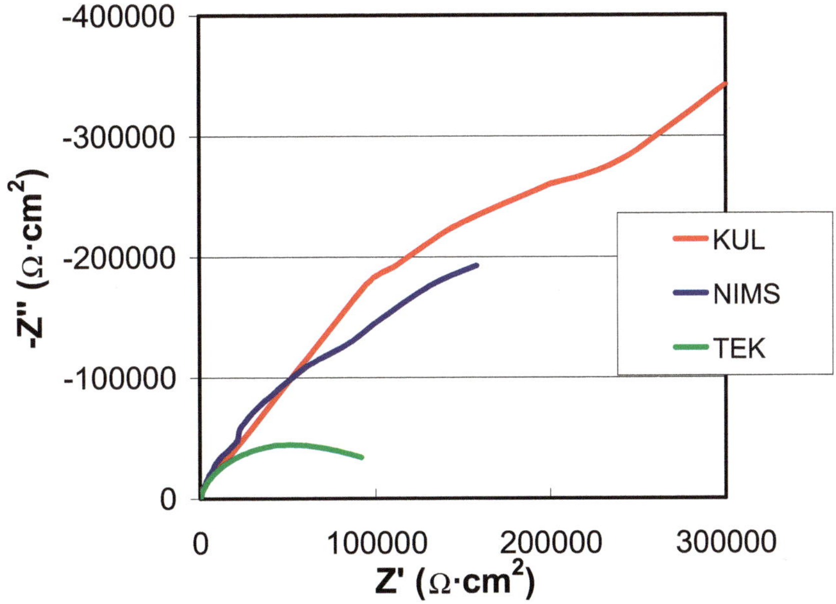

(a)

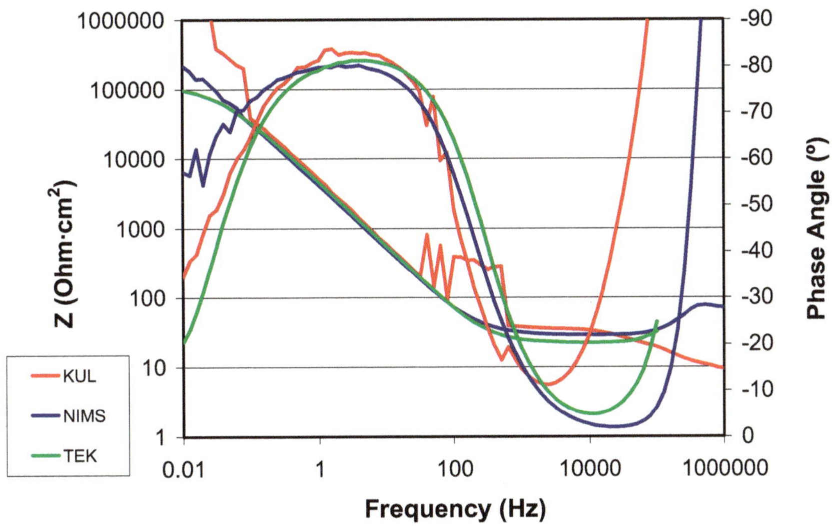

(b)

4.5 Nyquist (a) and Bode (b) plots obtained in PBS solution at OCP by three different laboratories

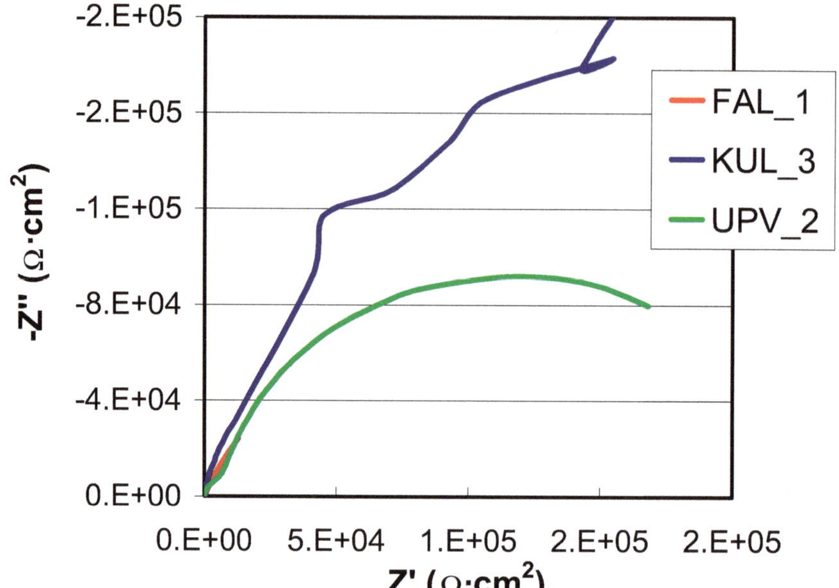

(a)

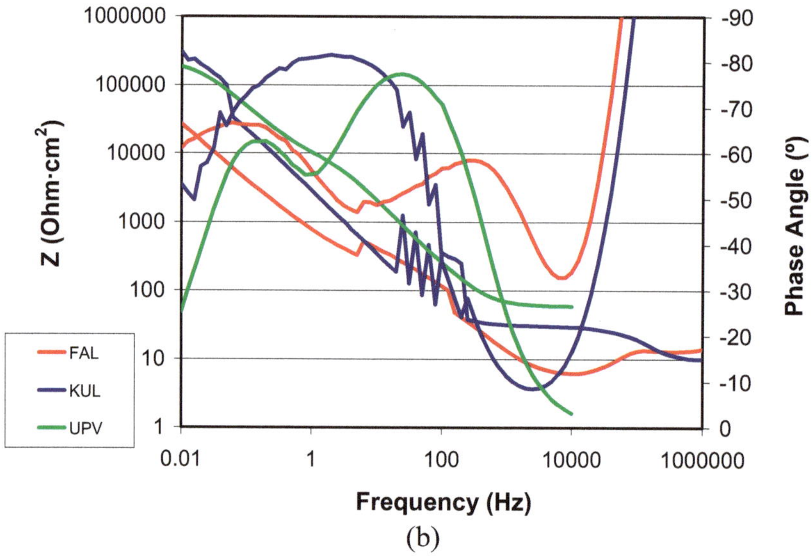

(b)

4.6 Nyquist (a) and Bode (b) plots obtained in PBS+Albumin solution at OCP by three different laboratories

i_p values. The first group corresponds to two laboratories which measured high negative current densities; the second group includes two laboratories which measured high positive current densities with high scatter in their measurements. Finally, the third group, which includes most laboratories, measured an average i_p value of 0.2 μA/cm². Note that Laboratory 5 reported using a coarse current range, which probably explains their high i_p value.

Table 4.4 Statistical evaluation of results obtained by each laboratory for the PBS and PBS+Albumin solutions at OCP

	PBS			PBS+Albumin		
	OCP (mV)	**R_s (Ω cm²)**	**R_p (kΩ cm²)**	**OCP (mV)**	**R_s (Ω cm²)**	**R_p (kΩ cm²)**
Lab1	–194±22	59±32	67±5	–263±32	62±12	64±4
Lab2	–200±30	33±2	130±30	–250±10	26±6	210±30
Lab3						
Lab4	–130±10	23±4	311±64	–221±20	34±5	235±15
Lab5	–305±105	24±2	36±24	–370±8	44±1	118±16
Lab6	672±105	4±0.6	2±0.7	96±237	–4±0.6	16±20
Lab7	–194±24	24±1	238±45	–246±30	28±2	322±43
Lab8	–130±20	24±9	169±14	–173±12	27±9	152±24
Lab9	–12±37	–2±0.8	470±368	–302±10	–1±0.6	5±4
Lab10	–214±19	37±5	440±15	–273±30	33±2	240±39
Lab11	–63±105	24±1	123±44	–468±12	19±4	73±14
Lab12	–214±48			–226±49		
Lab13	–363±12	21±2	56±21			
Lab14	–210±11	42±1	248±58	–209±21	43±1	289±90
Lab15	5±33	27±2	231±100	91±43	55±4	240±43
MV ± σ	**–100±247**	**26±15**	**194±148**	**–216±157**	**29±19**	**164±107**
MV ± σ ISO 5725:1994	**–157±113**	**23±13**	**186±122**	**–273±83**	**152±104**	
Accepted labs	**13**	**11**	**11**	**11**	**9**	**11**

R_p values show a reasonable repeatability within the individual laboratories, but a low reproducibility is observed among different laboratories. The scatter is reduced in the PBS+Albumin solution.

4.4 Experiment 3

Analogously to Experiment 2, this experiment was developed to analyse the capability of impedance measurements for characterising the biomaterial/electrolyte interface in biological media. Figures 4.5 and 4.6 show the Nyquist diagrams and Bode plots obtained at OCP in the PBS and PBS+Albumin solutions, respectively.

Parameters extracted by the participants from the EIS diagrams in PBS and PBS+Albumin are listed in Table 4.4.

OCP is a measured parameter, which means that no manipulation was made by the participants to obtain this value. In this case, only one of the laboratories measured an out-of-range value, which also corresponds to the laboratory which determined a negative i_p (Table 4.4).

Most of the laboratories observed negative OCP values. Laboratories 6 and 15 reported positive values: interestingly, both laboratories also reported problem with the evacuation of gas bubbles formed during cathodic cleaning. R_s values at the OCP are similar to those observed under applied potential, while R_p values are slightly lower at the OCP.

Qualitatively, the majority of laboratories found the same EIS spectra features at the OCP as at 0.15 V_{SHE}: CPE behaviour, solution resistance above 1 kHz, noise and interference above 20 kHz. The phase angle features (Figs. 4.5b and 4.6b) of most laboratories indicate the presence of one time constant while only three laboratories (Laboratories 6, 11 and 15) indicate the presence of two time constants. Laboratory 6 also observed more time constants in the PBS solution.

Some participants observed noise at low frequency, mainly at the OCP.

5
Discussion

The results obtained from the participating laboratories are analysed in this chapter with particular emphasis on repeatability (precision within an individual laboratory), reproducibility (precision among all involved laboratories), and critical appraisal of the procedures used to extract relevant data. Beyond this statistical interpretation, the physical (corrosion mechanisms) and clinical (metal release rate into the body) meaning is discussed. Finally, possible points to be considered for improving protocols are addressed.

5.1 Repeatability

The scatter obtained among single laboratories is compared in Table 5.1 with the results obtained in a similar round robin study concerning electrochemical corrosion and tribocorrosion measurements (CEFRACOR, [33]). Note that the system investigated in the latter round robin was quite different and involved a stainless steel immersed in sulphuric acid.

For potential (E_{corr}, E_b, OCP) measurements, the scatter range varied from a few millivolts (all round robins) up to 20–40 mV (CEFRACOR) or to 60 to 150 mV for the present COST experiments. This indicates that it was easier to achieve repeatability in potential measurements in a strong electrolyte such as sulphuric acid than in the neutral solution used as simulated body fluids.

The repeatabilities in i_{pp} values scatter from a very low value of approximately 0.2 μA/cm^2 (all RR) up to 7 μA/cm^2 for the present experiments and to 20 μA/cm^2 for the CEFRACOR round robin. Note that the average i_{pp} value was very similar in all experiments (9 to 16 μA/cm^2). So, unlike potential measurements, easier repeatability of passive current was obtained with the CoCrMo alloy in simulated body fluid compared to stainless steel in sulphuric acid. Nevertheless, the scattering in laboratory repeatability can be considered as similar in both round robins.

Table 5.2 shows the scatter in repeatability of the passivation current densities i_p and the OCP values obtained before EIS measurement after 1 h stabilisation. The scatter range is similar to that observed in the polarisation curves. Interestingly, in the latter tests, stabilisation of OCP lasted for only 10 min. This suggests that longer stabilisation times do not necessarily bring better repeatability in DC measurements.

Absolute repeatability in solution resistance lies in the range of a few tens of ohms, a value several orders of magnitude lower compared to the scatter in polarisation resistance (kΩ range). Note that better repeatability is observed at the OCP compared to the tests at applied potential. However, the relative scatters in R_s and in R_p are very similar, typically 5 to 30%. As in the case of DC measurements, some laboratories obtained good repeatability while others had more problems. In general, the same repeatability was observed for PBS and PBS+Albumin solutions and this for potentiodynamic, potentiostatic and EIS results.

Table 5.1 Minimum and maximum values of the standard deviation of the electrochemical parameters extracted from the potentiodynamic curves by the participating laboratories and CEFRACOR results

	COST533 (PBS solution)	COST533 (PBS+Albumin solution)	CEFRACOR
E_{corr} mV	±5 to ±139	± 0 to ± 150	± 4 to ± 17
i_{corr} mA/cm²	±0.00002 to ± 0.004	±0.00002 to ± 0.04	
i_{pp} mA/cm²	±0.0002 to ± 0.0072	0.0003 to ± 0.006	±0.0001 to ± 0.02
E_b mV	± 3 to ± 120	± 1 to ± 78	± 2 to ± 39
OCP mV	± 8 to ± 100	± 1 to ± 60	± 1 to ± 27

Table 5.2 Minimum and maximum values of the standard deviation of the electrochemical parameters extracted from the EIS tests by the participating laboratories

	COST533 (PBS solution)	COST533 (PBS+Albumin solution)
i_p mA/cm²	±0.000008 to ± 0.015	0.00002 to ± 0.007
OCP mV	± 10 to ± 105	± 8 to ± 49
R_s at OCP Ω	± 1 to ± 32	± 1 to ± 12
R_s at 0.15 V Ω	± 1 to ± 15	± 1 to ± 10
R_p at OCP Ω	± 5000 to ± 100 000	± 4000 to ± 90 000
R_p at 0.15 V Ω	± 8000 to ± 609 000	± 1000 to ± 443 000

It is somehow striking to observe the scatter in repeatability among laboratories with some laboratories obtaining insignificant dispersion while others observed large scatter. One of the reasons was that some laboratories repeated the experiments more than three times and extracted the best three values for statistical evaluation. Some other laboratories kept strictly to the protocol and repeated the experiments only three times without excluding any outliers from their statistics. In this latter case, repeatability was of course worse. To avoid such problems, in future round robins, the desired number of repetitions should be clearly stated and imposed on all participants.

Other sources of lack of repeatability are related to statistical variations in sample preparation (surface roughness and contamination, local variations in metal composition), and/or solution preparation (oxygen content, water contaminants). Electrochemical processes are highly sensitive to very small variations in surface state and therefore some variations are to be expected. Other uncertainties are associated with errors in the extraction procedure of data from the polarisation curves. The fact that similar scatter ranges are observed for extracted data (E_{corr} and E_b) and measured data (OCP) indicates, however, that the proposed extraction procedure was sufficiently accurate.

Furthermore, lack of repeatability can be due to random shortcomings in the instrumentation and/or in the electrode set-up of the electrochemical cell. These points could be checked by using appropriate calibration procedures involving well defined dummy cells (instrumentation) and ohmic drop effects (position of the reference electrode with respect to the working electrode).

Table 5.3 Average values and standard deviation of the electrochemical parameters extracted from potentiodynamic curves and EIS by the individual laboratories. Comparison with EFC and CEFRACOR results

	COST533 (PBS solution)	COST533 (PBS+Albumin solution)	EFC	CEFRACOR
E_{corr} mV	−363 ± 319	−510 ± 162	−306 ± 20	−202 ± 26
i_{corr} mR/cm²	0.0026 ± 0.002	0.0031 ± 0.0023		
i_{pp} mA/cm²	0.0074 ± 0.0021	0.0081 ± 0.0029	0.004 ± 0.0015	0.016 ± 0.008
E_b mV	728 ± 41	710 ± 55		1089 ± 58
i_p mA/cm²	0.0002 ± 0.0001	0.0004 ± 0.0002		0.00011 ± 0.00012
OCP_{EIS} mV	−157 ± 113	−273 ± 83		−169 ± 81
R_s 0.15 V kΩ	30 ± 11	35 ± 12	± 5.5	
R_p 0.15 V kΩ	604 ± 399	517 ± 280	± 383	
R_s OCP Ω	23 ± 13	32 ± 11	± 4.8	
R_p OCP kΩ	186 ± 122	152 ± 104	± 60	

5.2 Reproducibility

The average values of the data extracted by the individual laboratories are listed in Table 5.3 together with corresponding values obtained in the EFC [34] and CEFRACOR [33] round robins. Similar reproducibility in the current values as well as in the E_b and OCP data is found in all round robin results. Larger scatter in E_{corr} is observed in the present COST investigation, which suggests that the electrochemical conditions established at E_{corr} are less well defined in neutral solution than in acidic solutions. This point will be further discussed in Section 5.4. The absolute difference in E_b potential in the pH 7.4 solution (present COST round robin) and the pH 0.5 acid solution (EFC and CEFRACOR round robins) corresponds well with the change in reversible potential of the transpassive reactions (oxidation of water, oxidation of Cr to Cr^{6+}) that diminishes at 37°C by 0.061 V for each pH unit.

The reproducibility of R_s and R_p (standard deviation of the average values) is higher than the average repeatability found within individual laboratories. This is to be expected since the reproducibility not only reflects the repeatability of the procedure but also possible differences in instrumentation, cell design and extraction procedures among the participating laboratories.

Albumin has no significant effect on any of the parameters listed in Table 5.3 except the passivation current density measured before starting the EIS measurement at imposed potential (i_p). Albumin increases this value by a factor of 2.

5.3 Extraction procedures

One recurrent question is how far differences in data extraction procedures affect the scatter of the results. The corrosion current density is usually determined by the intercept of the extrapolated cathodic and anodic linear parts of the log $|i|$ vs E plots. Most modern computer controlled potentiostats provide integrated software routines to calculate this. In order to evaluate the reliability of such routines, corrosion current densities were extracted from selected polarisation curves (measured in PBS) by the coordinators. Note that the coordinators worked independently and that coordinator 1 was not informed about which laboratory performed the measurement. The procedure followed by coordinator 1 consisted of first identifying the linear domains in the log |*i*| vs *E* plots situated at least 100 mV below (cathodic) or above (anodic) the corrosion potential. Linear domains not within the potential range ±300 mV with respect to E_{corr} were not considered. As a successive step, lines were drawn manually and the log |*i*| value was graphically determined from the intercept and converted into i_{corr} values. Coordinator 2 followed the same procedure but only considered potentials between 50 and 100 mV above (anodic) or below (cathodic) E_{corr}. The difference in procedure reflects the particular experience of each laboratory. The data obtained are presented in Table 5.4.

Interestingly, the coordinators were not able to identify suitable linear domains in 2 out of 5 analysed polarisation curves and this despite the fact that the participants supplied corrosion current densities. As an example of this, Fig. 5.1 presents two

Table 5.4 Corrosion current density values extracted by several laboratories and coordinators

	i_{corr} (μA/cm²)		
	Coordinator 1	Coordinator 2	Participant
Laboratory A	2.34	1.9	1.88
Laboratory B	nd	3.4	4.1
Laboratory C	4.36	2.28	6.3
Laboratory D	3.98	ND	0.01
Laboratory E	ND	ND	2.71

ND, not determined because no linear domain could be found.

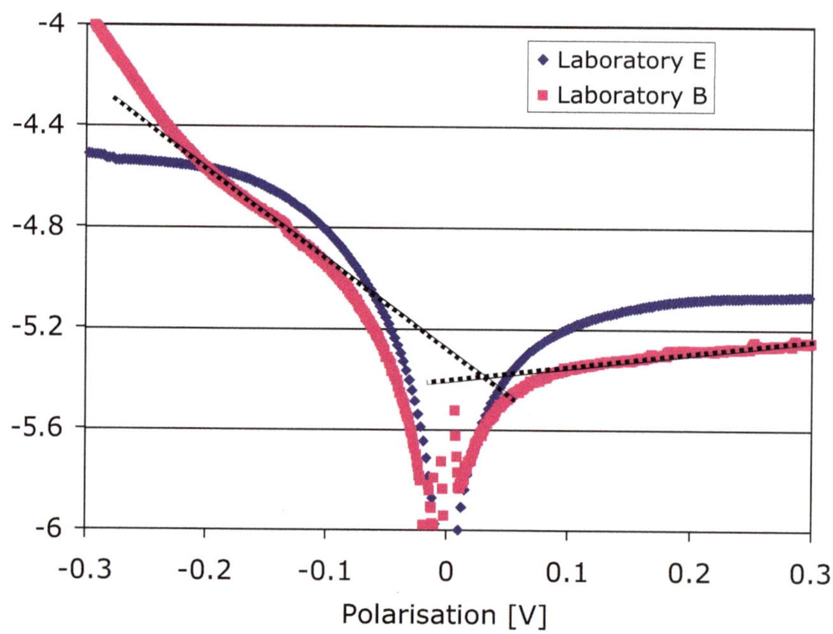

5.1 Polarisation curves of two different laboratories with (Laboratory B) and without (Laboratory E) linear domain

polarisation curves from Laboratory B (linear domains visible) and from Laboratory E (no linear domains).

In two cases (Laboratories A and C), coordinators and participants reached consistent values within the repeatability range. Strong disagreements were found in all other cases where either none of the coordinators (Laboratory E) or only one of the coordinators (Laboratory B and D) could find satisfactory linear domains. These results clearly demonstrate the need for an improved unified extraction procedure for i_{corr} values from polarisation curves. Furthermore, the fact that i_{corr} values were extrapolated from curves not showing any exploitable linear domain should raise awareness that automated computer routines should be used carefully and only after a visual check of the appropriateness of the data set.

All laboratories found impedance spectra showing a non-perfect RC behaviour with only one apparent time constant. The Nyquist plots show depressed semicircles due to imperfect surfaces. In order to check the effect of the extraction procedure for R_p and R_s carried out by each individual laboratory, the coordinators fitted selected impedance spectra by using a simple RC equivalent circuit (with R corresponding to R_p) in series with a resistance (R_s). The results obtained are compared in Table 5.5 to the values supplied by the participants. Generally, a good correlation was found between coordinators and participants. Only in the case of Laboratory E were the coordinators' values for R_s and R_p systematically higher. Based on this limited set of results, one can conclude that the extraction procedure is not a key factor affecting the reliability of EIS data.

5.4 Physical interpretation of the measurements

Polarisation curves are usually used to identify the reactions determining the electrochemical response and the corrosion behaviour. All laboratories found a very similar shape for the polarisation curves. The transpassive potential was clearly identified above approximately 0.7 V characterised by an increase in current due to water oxidation and/or oxidation of chromium to hexavalent species. A well defined passive plateau was also identified below the transpassive potential. At very low potentials, all laboratories observed strong currents due to the reduction in water. The polarisation behaviour in the potential domain close to the corrosion potential shows significant scatter not only among laboratories but even within individual laboratories. This is not surprising since the currents in this range are usually low and are determined by several cathodic and anodic reactions.

Table 5.5 Polarisation resistance values extracted by several laboratories and coordinators

	R_s ($\Omega\cdot cm^2$)		R_p ($k\Omega\cdot cm^2$)	
	Coordinators	Participant	Coordinators	Participant
Laboratory A	58	58	1337	1330
Laboratory B	32	31	457	466
Laboratory C	38	43	610	690
Laboratory D	31	34	1360	1360
Laboratory E	35	9.6	1033	550

In the present systems, some doubts exist about the electrochemical behaviour of the alloy components. According to Hodgson [11], cobalt is expected to be in the active state while Cr is expected to be passive. The exact state of the alloy during cathodic polarisation and at potentials close to the corrosion potential is therefore not well defined and may vary depending on the nature of the native oxide film covering the surface before immersion. Two cathodic reactions are expected to play a role: the reduction of water and the reduction of oxygen. The former is expected to contribute little for relatively high potentials close to the corrosion potential. The kinetics of oxygen reduction depends very much on the presence or absence of a passive film on the electrode surface that usually significantly inhibits the charge transfer kinetics of oxygen. In the absence of passive films, the reaction is usually found limited by mass transport due to the low concentration of oxygen. Mass transport limitation manifests itself by a current plateau of magnitude between 50 and 100 $\mu A/cm^2$. Indeed, such plateaux are observed systematically by some laboratories and in a non-systematic way by other laboratories. Some laboratories do not observe such plateaux. The reasons for these discrepancies are not well understood: they may be related to the presence of residual oxide films or to the poor control of the oxygen content in the test solution. The poor control of the cathodic reactions and of the surface state may also explain the fact that no systematic Tafel behaviour is observed in the polarisation curves. Clearly, the elucidation of the exact causes of the scatter in results close to the corrosion potentials would require a better understanding of the surface chemical and electrochemical reactions involved. However, this lies beyond the scope of the present work. Suffice it here to recommend addressing these points in an improved protocol.

Polarisation curves are widely used to evaluate the effect of selected parameters, such as the presence of albumin on the corrosion behaviour. The main influence of albumin observed here on the electrochemical polarisation behaviour consists of the decrease in OCP and E_{corr} and in the increase in i_{corr}. Such influences have already been observed by several authors and interpreted as resulting from adsorption effects that inhibit the cathodic reaction. Interestingly, no clear effect of albumin is observed on the electrochemical kinetics measured by EIS or by polarisation in the passive domain. Indeed, neither R_p nor i_{pp} values are systematically affected by the presence of albumin. This is consistent with a cathodic inhibitor effect of albumin [20].

5.5 Comments on precision with respect to clinical applications

Corrosion rate (V_{corr}) is of particular interest for clinical applications since it determines the amount of metal released into the body as oxidised species. Two electrochemical methods are usually used to determine the corrosion rate: the first consists of extracting the corrosion current density i_{corr} from polarisation curves and the second consists of measuring the polarisation resistance R_p using EIS.

The polarisation resistance can be converted into a corrosion current density using equation 5.1

$$i_{corr,\,Rp} = \beta_a\,\beta_c\,(\beta_a + \beta_c)^{-1}\,R_p^{-1} \tag{5.1}$$

where β_a and β_c correspond to the Tafel coefficients of the metal oxidation and of the reduction in oxidiser. Tafel coefficients typically lie in the range 20 to 50 mV.

Corrosion rates can be expressed in terms of material release per unit surface area and unit time and of depth corroded per unit time. The conversion of corrosion current density into corrosion rate is given by equations 5.2 and 5.3

$$V_{corr} = 32.704\ i_{corr}\ M\ n^{-1} \qquad \text{mg/dm}^2\ \text{year} \qquad [5.2]$$

$$V_{corr} = 3.27\ i_{corr}\ M\ n^{-1}\ \rho^{-1} \qquad \mu\text{m/year} \qquad [5.3]$$

where i_{corr} is expressed in μA/cm², M is the atomic mass in g/mol, n is the charge number (dimensionless) and ρ is the density in g/cm³. For a coarse estimation of corrosion rates, we considered the atomic mass and density of pure cobalt (58.9 g/mol and 8.9 g/cm³, respectively) and a charge number of 2.3 (3 for chromium, 2 for cobalt).

By converting the i_{corr} values extracted from polarisation curves (values excluding outliers according to ISO5725) into corrosion rates (Table 5.6), one obtains release rates of 2177 ± 1675 mg/dm² year and 2596 ± 1926 mg/dm² year for the PBS and PBS + albumin solution, respectively. For simplicity, considering that the corroding area of an implant corresponds to 1 dm² and supposing the same corrosivity of body fluids as the PBS solution, one would obtain a theoretical release rate of metal into the body of 6 mg/day or 2200 mg/year. Undoubtedly, such a massive material release would have a tremendous impact on body response, taking into account that the average free metal content of an adult is approximately 5 mg.

In-vivo corrosion rates of 22 mg/year have been reported in the past [2]. These in-vivo rates are much lower than the corrosion rates obtained in this study. Note that the in-vivo rates were estimated from the analysis of the metal content in selected tissues or body fluids and thus reflect the local accumulation rate rather than the effective release rate from the implant. In addition, it is likely that body fluids are less corrosive due to the presence of a large amount of organic molecules that can adsorb and block the metal surface.

Table 5.6 Corrosion rates (mg/dm² year) calculated from DC results

	PBS	PBS+Albumin
Lab 1	583±170	846±324
Lab 2	25±124	27±16
Lab 3		
Lab 4	3657±193	3712±1013
Lab 5	3394±831	1082±607
Lab 6	3776±3505	3297±1253
Lab 7		
Lab 8	1949±452	1283±372
Lab 9	2490±2003	5290±982
Lab 10	52 369±192	5367±125
Lab 11	29 337±15 917	32 199±33 761
Lab 12		
Lab 13	1409±380	
Lab 14	224±18	379±284
Lab 15	1286±333	2141±407
MV ± σ ISO 5725:1994	**2177±1675**	**2596±1926**
Accepted labs	**8/12**	**9/12**

By taking the repeatability as a measure of precision, one obtains a precision ranging from 16 mg/dm^2 year (0.02 μA/cm^2) to 3350 mg/dm^2 year (4 μA/cm^2) for the corrosion rates determined from polarisation curves. Clearly, even in the best case, the precision of polarisation curves for the determination of corrosion rates seems to be largely insufficient for practical application, where much more accurate measurements are probably needed to assess the impact of corrosion on the biological response. A significant improvement in precision is needed to characterise quantitatively corrosion rates of CoCrMo alloys in biological fluids.

Corrosion current densities can be calculated from measured polarisation resistance using equation 5.1 and converted into corrosion rates using equations 5.2 or 5.3. Table 5.7 lists the corrosion rates obtained through this conversion of polarisation resistances measured in the PBS and PBS+Albumin solutions at OCP and imposed potential using EIS. Average corrosion rates obtained through EIS at OCP are approximately 100 to 125 mg/dm^2 year (with no significant effect of albumin), a value one order of magnitude lower than the corrosion rates extracted from polarisation curves. This difference can be explained by the different electrochemical conditions established in the two tests. EIS is carried out on passive surfaces while the extraction of the corrosion rate from polarisation curves is carried out in a potential domain where the surface is not necessarily passive (see next section) and may corrode in a more severe active mode.

Improved precision of corrosion rates is found using EIS measurements where the average repeatability at OCP in the PBS solution ranges from 3 to 300 mg/dm^2 year depending on the laboratory. Slightly better repeatability is found in the PBS+Albumin solution but here, however, a larger number of outliers is found. EIS measurements at imposed potential exhibit slightly better repeatability and reproducibility. The

Table 5.7 Corrosion rates (mg/dm^2 year) calculated from impedance results

	PBS		PBS+Albumin	
	OCP	Passive	OCP	Passive
Lab 1	115±26	157±9	327±21	154±7
Lab 2	110±29	19±8	101±14	24±4
Lab 3				
Lab 4	69±16	41±5	89±5	30±0.4
Lab 5	969±933	222±100	181±32	52±6
Lab 6	10 498±3485	1388±1202	3113±2286	1043±646
Lab 7	91±18	21±4	66±10	30±13
Lab 8	125±14	34±3	140±20	26±2
Lab 9	64±37	664±385	75 050±131 015	5488±8931
Lab 10	47±1	28±4	89±16	30±0.3
Lab 11	159±56	19±3	291±51	32±3
Lab 12				
Lab 13	411±139	8±2		
Lab 14	87±18	29±4	76±20	30±3
Lab 15	107±58	22±7	89±14	310±62
MV ± σ	**875±2722**	**177±456**	**6926±38 230**	**604±2625**
MV ± σ ISO 5725:1994	**125±79**	**25±9**	**104±38**	**29±3**
Accepted labs	**10/13**	**9/13**	**8/12**	**6/12**

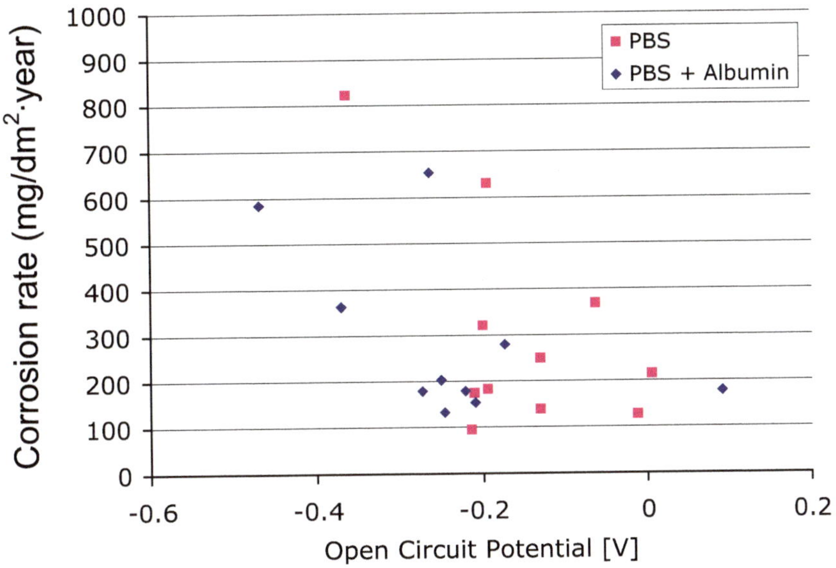

5.2 Corrosion rate versus applied OCP in experiment 3

reason for this difference may lie in the fact that corrosion processes are electrochemical in nature and therefore, the reaction rate depends on the prevailing electrode potential. In Fig. 5.2, the corrosion rates determined from the results of each individual laboratory (without outliers) are plotted against the corresponding measured OCP potential. The data show large scatter but a trend of smaller corrosion rate with increasing potential is clearly visible. Figure 5.2 confirms that the electrode potential is one of the critical factors affecting the reaction rate. Any prediction of the corrosion rate of CoCrMo alloys therefore requires the identification of the electrode potential established between the implant and the surrounding biological environment. Such a value depends, in principle, on several factors such as the pH and oxygen content of the body fluids as well as their temperature and flow conditions. At present, such factors are not well defined in biological environments. Clearly, their quantification is a prerequisite for a correct and reliable estimation of corrosion rates of alloys used for biomedical implants.

5.6 Improvements in experimental protocols and data reporting

Several factors have been identified as responsible for increasing the scatter in the measurements and calculated parameters; therefore, improvements in the experimental procedure should be considered for correct experimental practice and for reporting scientific results:

- ✓ Surface finishing: Exact specification of the polishing method (including the specific lubricant to be used for polishing). Better description of surface preparation should be reported in publications. Passivation in strongly oxidising environments (for example, nitric acid) could be evaluated as a method to produce reproducible surfaces.

- ✓ Solution conditions: Aeration of the solution; oxidant access to the working electrode surface may depend on the electrochemical cell configuration; therefore, oxygen content should not be limited by aeration conditions. Possibilities such as introducing air into the solution during the electrochemical measurements should be considered in the improved protocol for better reproducibility of the data.
- ✓ Electrochemical pre-polarisation: Applied potential for the cathodic cleaning in the electrochemical pre-treatment should avoid hydrogen bubble generation. In the case of chemical passivation in nitric acid, cathodic cleaning could be suppressed or limited to a small cathodic polarisation at less negative potentials.
- ✓ Number of repetitions: Repeatability of the experimental results can only be analysed if the exact number of replicas to be carried out by the laboratory is specified; therefore, imposing the number of repetitions (no more, no less) of each test would improve the statistical analysis of the results.
- ✓ Data extraction: Specification of how to extract variables from measured data should be improved. Especially critical is the determination of the corrosion current density by the Tafel extrapolation and the polarisation resistance from impedance spectra (fitting procedures for calculating the electrical parameters should be described).
- ✓ Fundamental understanding: Mechanisms involved in the electrochemical reaction should be better understood for reducing the scatter of the measurements, both within individual laboratories and among different laboratories.

6
Guidelines

This interlaboratory investigation was initiated to obtain an overview of the precision of DC and AC electrochemical measurements used for corrosion investigation of biomedical alloys such as CoCrMo alloys. Another aim was to identify ways to improve test protocols and 'codes of practice' allowing for the comparison of results obtained by different laboratories in the field of biocorrosion. A third aim of this inter-laboratory study was to provide an overview of the network capabilities of laboratories involved in the electrochemical characterisation of biomedical alloys as a prerequisite for future joint research projects. The following conclusions can be drawn from this study:

- The '*Inter-laboratory study on electrochemical methods for characterisation of CoCrMo biomedical alloys in simulated body fluids*' was successfully completed by 15 different laboratories from Europe and Japan. All laboratories which declared their interest carried out the experiments within the planned timescale and broadly according to the proposed protocol. Some laboratories recognised limitations in their experimental set-up and defined ways to improve it. These facts attest to the capability and willingness of the participants to cooperate in network activities.
- In general, qualitative agreement has been found in the measurements. The results obtained were consistent with the literature on corrosion of CoCrMo alloys. Stationary and non-stationary techniques yielded consistent results.
- Repeatability and reproducibility are consistent with results from previous electrochemical interlaboratory studies on model systems. However, precision needs to be improved since the present state-of-the-art cannot resolve between negligible corrosion rates and hazardous rates of release of metals in the body.
- Critical factors affecting precision have been identified: surface preparation, chemical and electrochemical pre-conditioning, solution aeration and procedures for the extraction of relevant parameters. Thus, as good practice, such parameters need to be carefully described in future publications. Furthermore, they should be included in improved protocols.
- A better understanding is needed of the phenomena involved, in particular of the cathodic and anodic reaction mechanisms, as a prerequisite for more precise quantification.

Acknowledgements

The authors would like to thank the following for participation in this inter-laboratory investigation: J. de Damborenea and Maria Angeles Arena (CENIM), Jean Geringer (CIS), Pierre Ponthiaux and François Wagner (ECP), Patrik Schmutz and Olga Guseva (EMPA), Jelena Stojadinovic and Julien Perret (EPFL), Satish Achanta (FALEX), Bernard Normand (INSA), Ingrid Milosev and Tadeja Kosec (JIS), Jean-Pierre Celis and Animesh Basak (KUL), Sachiko Hiromoto (NIMS),

Amaia Igartua and Raquel Bayon (TEKNIKER), Takao Hanawa and Yusuke Tsutsumi (TMDU), Yu Yan (UL), Luís Augusto Rocha and Edith Ariza (UM), and Carlos Valero (UPV).

We would also like to thank COST 533 and EFC-WP18 for endorsing the initiative.

Also acknowledged are Dr C. Rieker, Zimmer Orthopaedics Europe Ltd, for his contribution in assessing the clinical implications and Surgival for supplying the CoCrMo alloy.

Appendix A
Direct current (DC) results: Polarisation curves with and without albumin obtained by each laboratory

A.1 Polarisation curves in PBS **without albumin**: 3 repeated tests for each laboratory

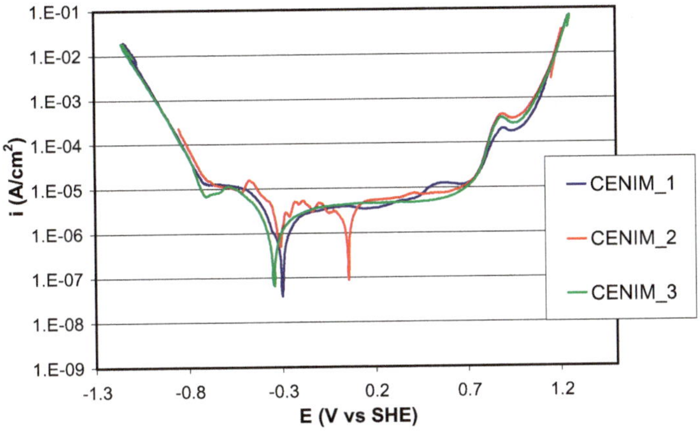

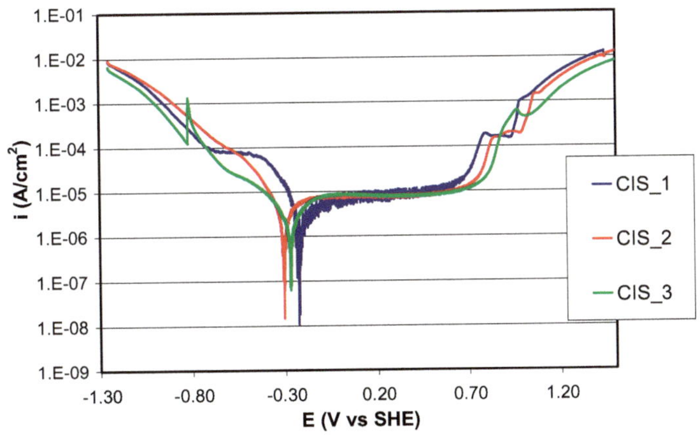

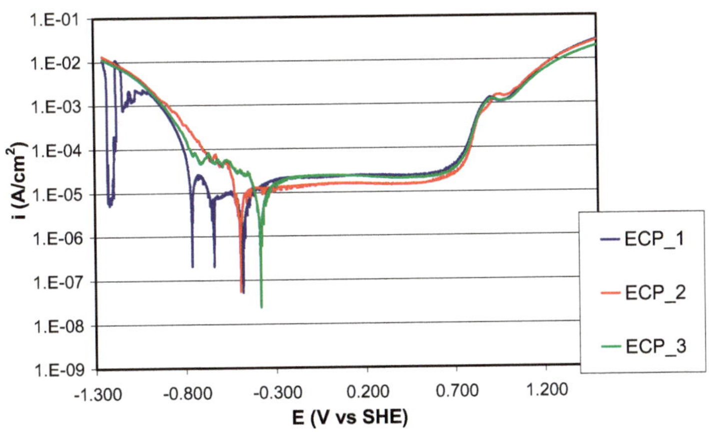

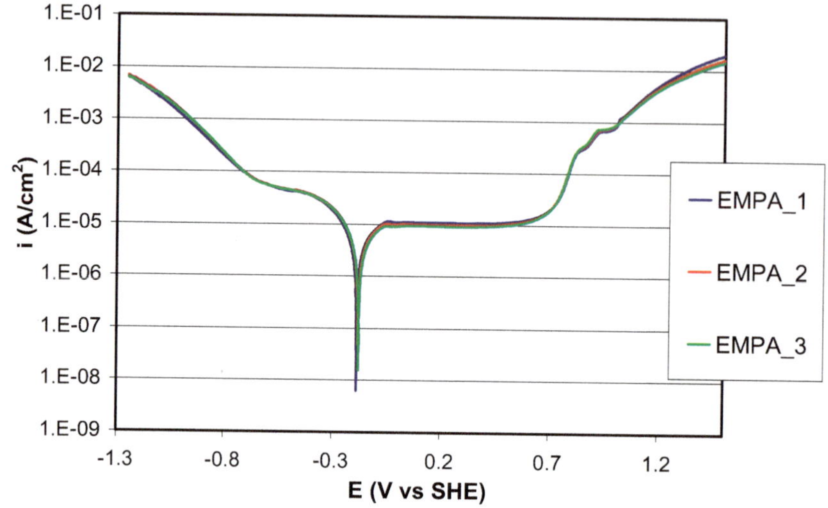
1.E-01
1.E-02
1.E-03
1.E-04
1.E-05
1.E-06
1.E-07
1.E-08
1.E-09
i (A/cm²)
-1.3
-0.8
-0.3
0.2
0.7
1.2
E (V vs SHE)
EMPA_1
EMPA_2
EMPA_3

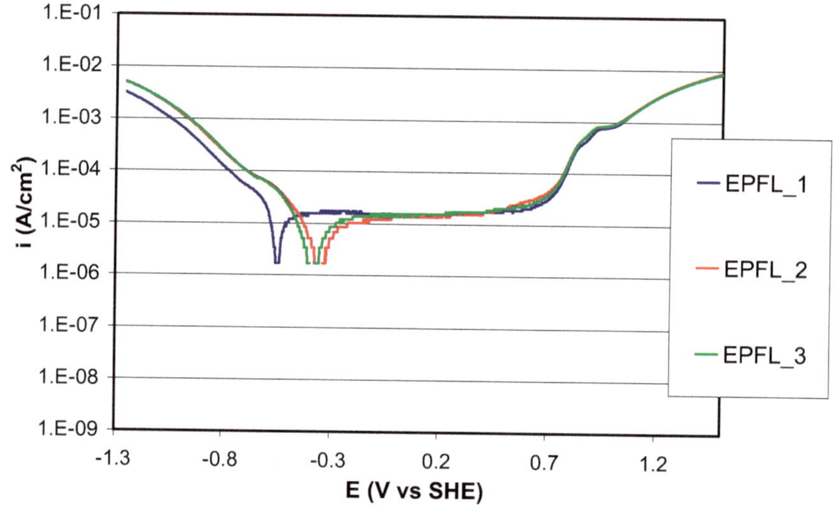
1.E-01
1.E-02
1.E-03
1.E-04
1.E-05
1.E-06
1.E-07
1.E-08
1.E-09
i (A/cm²)
-1.3
-0.8
-0.3
0.2
0.7
1.2
E (V vs SHE)
EPFL_1
EPFL_2
EPFL_3

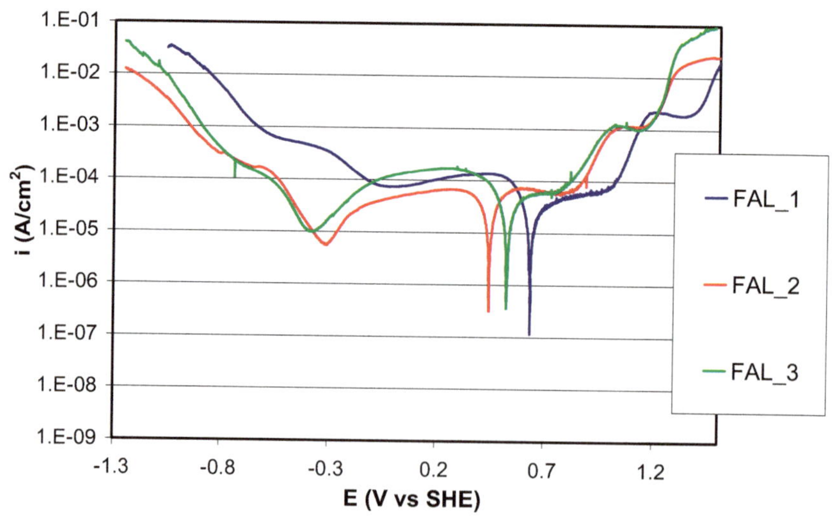
1.E-01
1.E-02
1.E-03
1.E-04
1.E-05
1.E-06
1.E-07
1.E-08
1.E-09
i (A/cm²)
-1.3
-0.8
-0.3
0.2
0.7
1.2
E (V vs SHE)
FAL_1
FAL_2
FAL_3

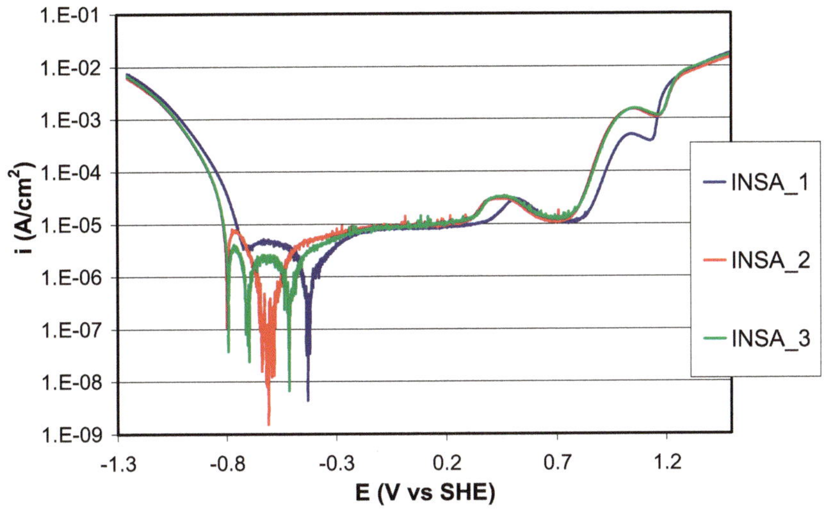

1.E-01
1.E-02
1.E-03
1.E-04
1.E-05
1.E-06
1.E-07
1.E-08
1.E-09
i (A/cm^2)
-1.3
-0.8
-0.3
0.2
0.7
1.2
E (V vs SHE)
INSA_1
INSA_2
INSA_3

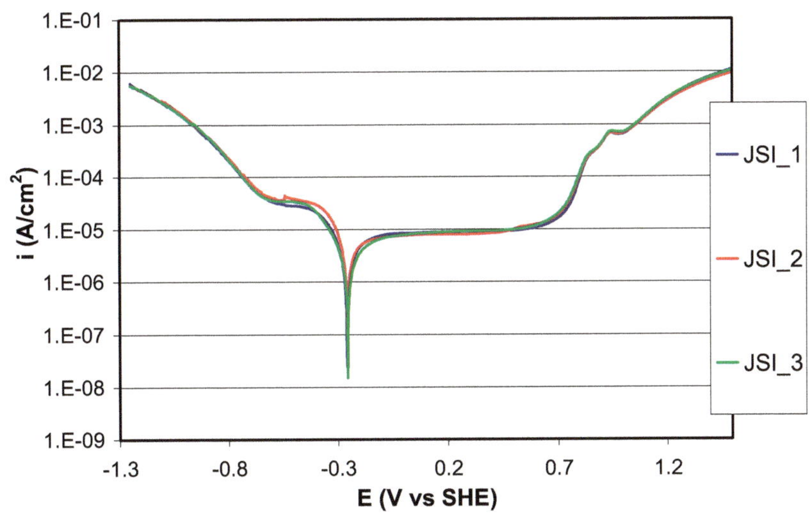

1.E-01
1.E-02
1.E-03
1.E-04
1.E-05
1.E-06
1.E-07
1.E-08
1.E-09
i (A/cm^2)
-1.3
-0.8
-0.3
0.2
0.7
1.2
E (V vs SHE)
JSI_1
JSI_2
JSI_3

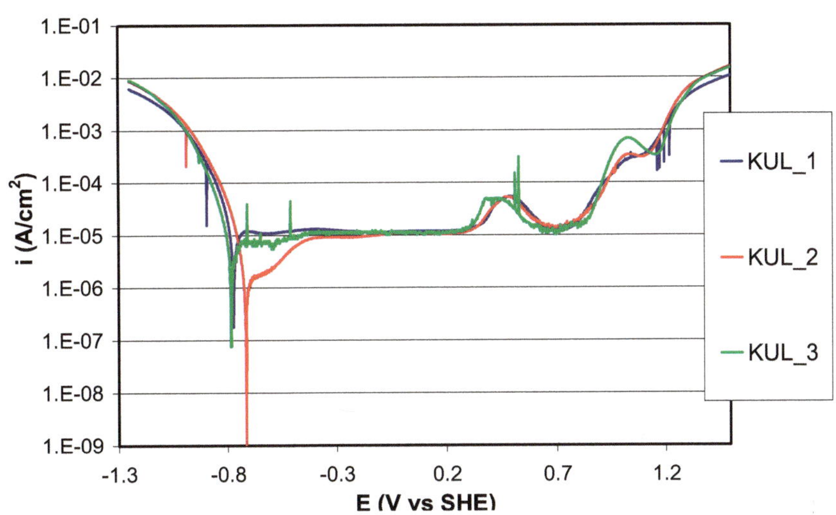

1.E-01
1.E-02
1.E-03
1.E-04
1.E-05
1.E-06
1.E-07
1.E-08
1.E-09
i (A/cm^2)
-1.3
-0.8
-0.3
0.2
0.7
1.2
E (V vs SHE)
KUL_1
KUL_2
KUL_3

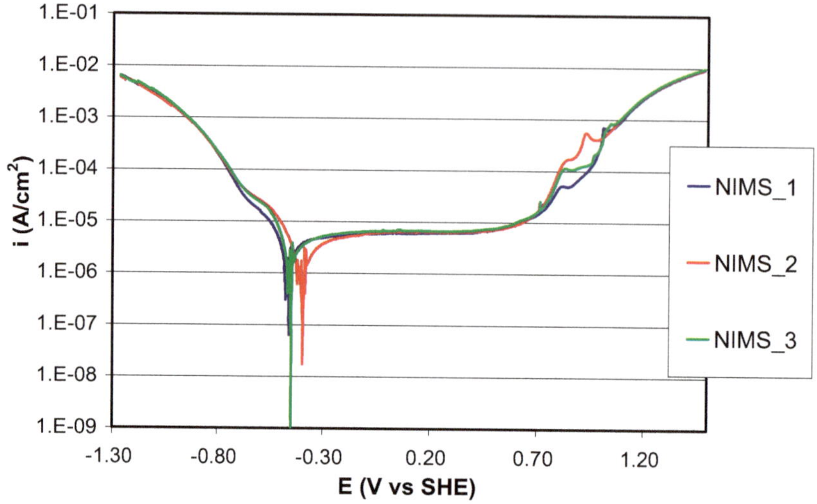
1.E-01
1.E-02
1.E-03
1.E-04
1.E-05
1.E-06
1.E-07
1.E-08
1.E-09
i (A/cm^2)
-1.30
-0.80
-0.30
0.20
0.70
1.20
E (V vs SHE)
NIMS_1
NIMS_2
NIMS_3

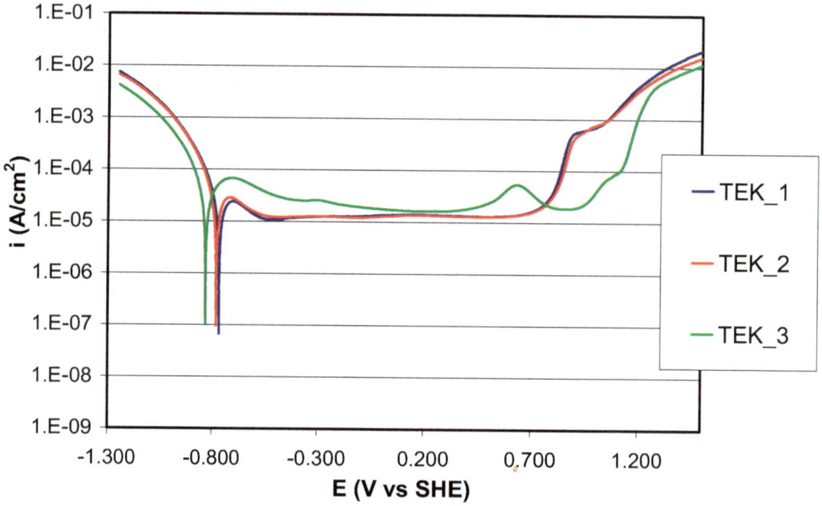
1.E-01
1.E-02
1.E-03
1.E-04
1.E-05
1.E-06
1.E-07
1.E-08
1.E-09
i (A/cm^2)
-1.300
-0.800
-0.300
0.200
0.700
1.200
E (V vs SHE)
TEK_1
TEK_2
TEK_3

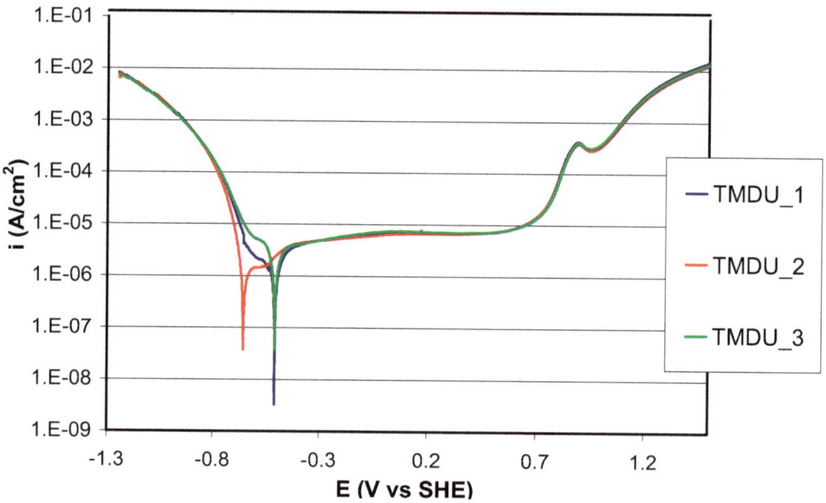
1.E-01
1.E-02
1.E-03
1.E-04
1.E-05
1.E-06
1.E-07
1.E-08
1.E-09
i (A/cm^2)
-1.3
-0.8
-0.3
0.2
0.7
1.2
E (V vs SHE)
TMDU_1
TMDU_2
TMDU_3

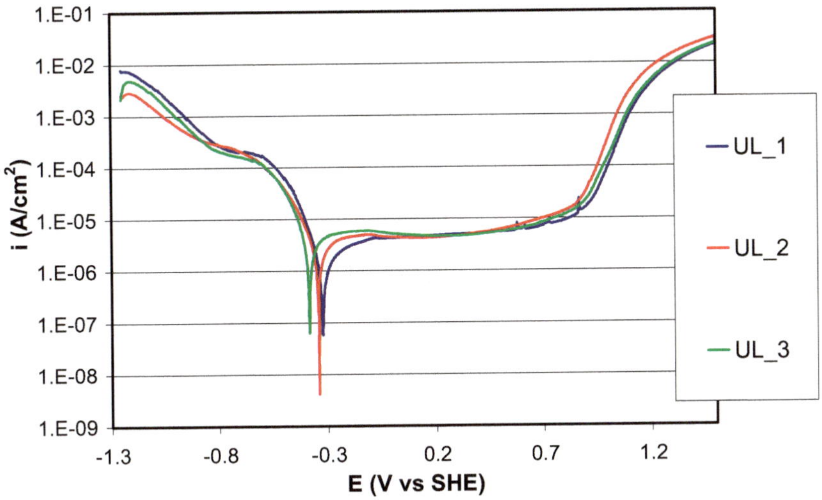
1.E-01
1.E-02
1.E-03
1.E-04
1.E-05
1.E-06
1.E-07
1.E-08
1.E-09
i (A/cm2)
-1.3
-0.8
-0.3
0.2
0.7
1.2
E (V vs SHE)
UL_1
UL_2
UL_3

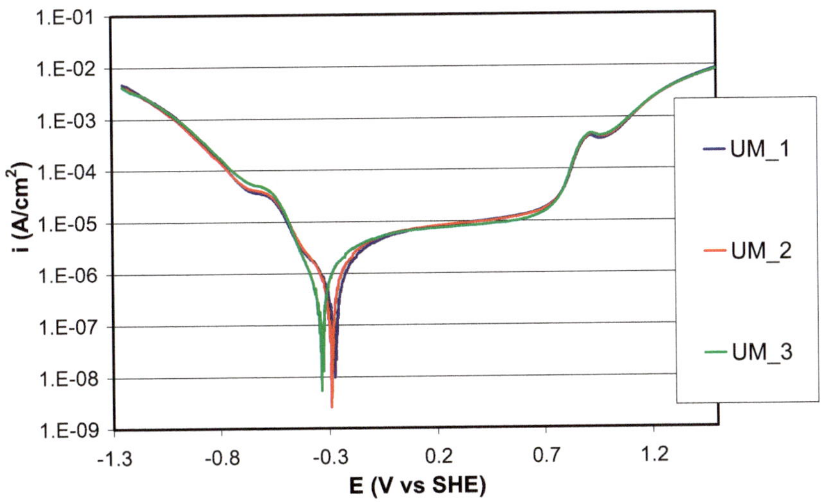
1.E-01
1.E-02
1.E-03
1.E-04
1.E-05
1.E-06
1.E-07
1.E-08
1.E-09
i (A/cm2)
-1.3
-0.8
-0.3
0.2
0.7
1.2
E (V vs SHE)
UM_1
UM_2
UM_3

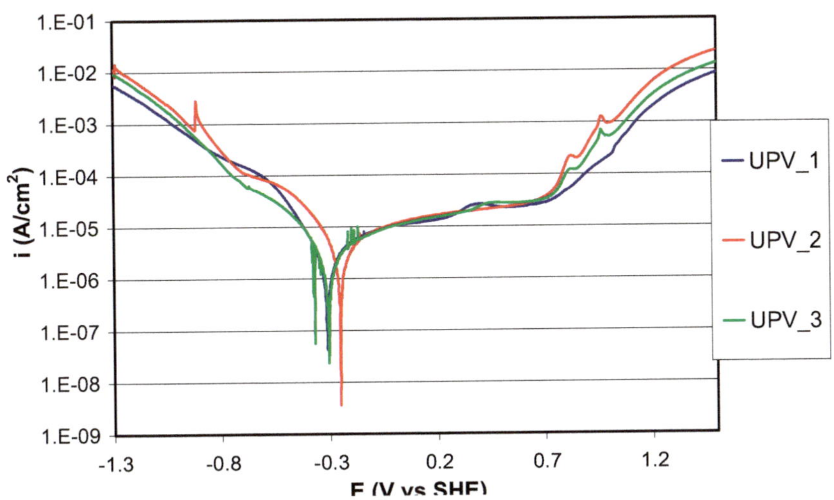
1.E-01
1.E-02
1.E-03
1.E-04
1.E-05
1.E-06
1.E-07
1.E-08
1.E-09
i (A/cm2)
-1.3
-0.8
-0.3
0.2
0.7
1.2
UPV_1
UPV_2
UPV_3

A.2 Polarisation curves in PBS **with albumin**: 3 repeated tests for each laboratory

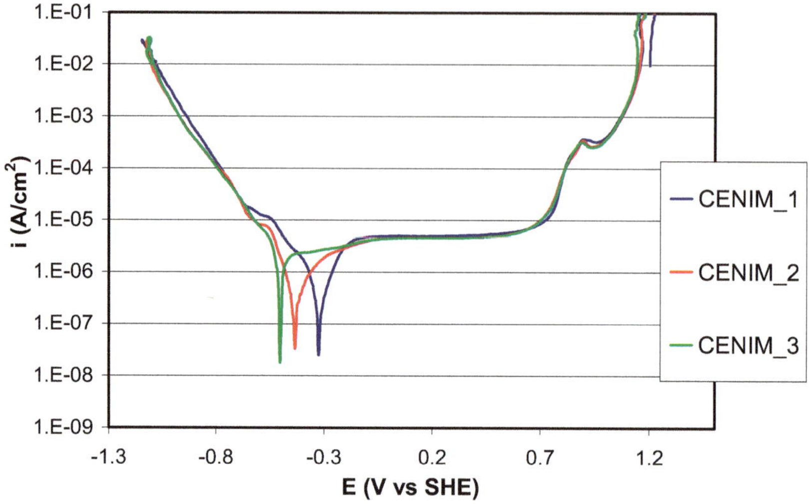

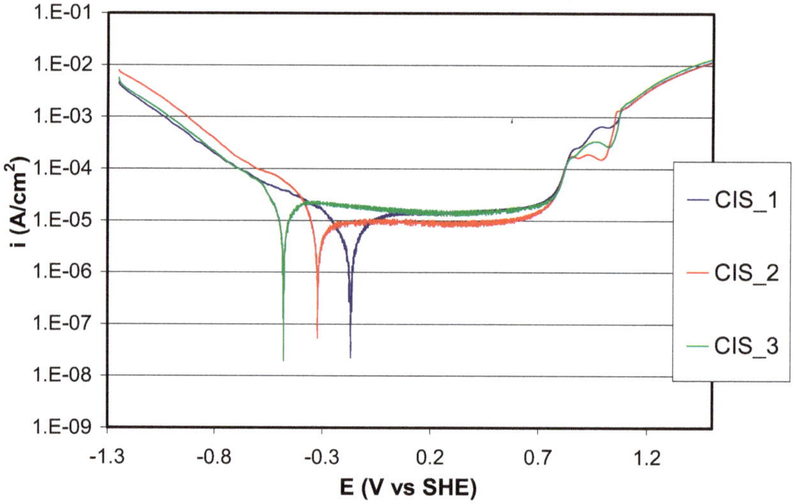

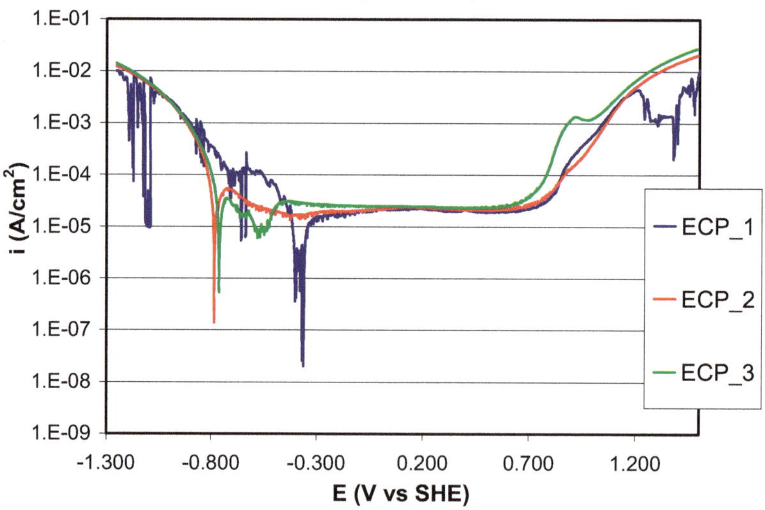

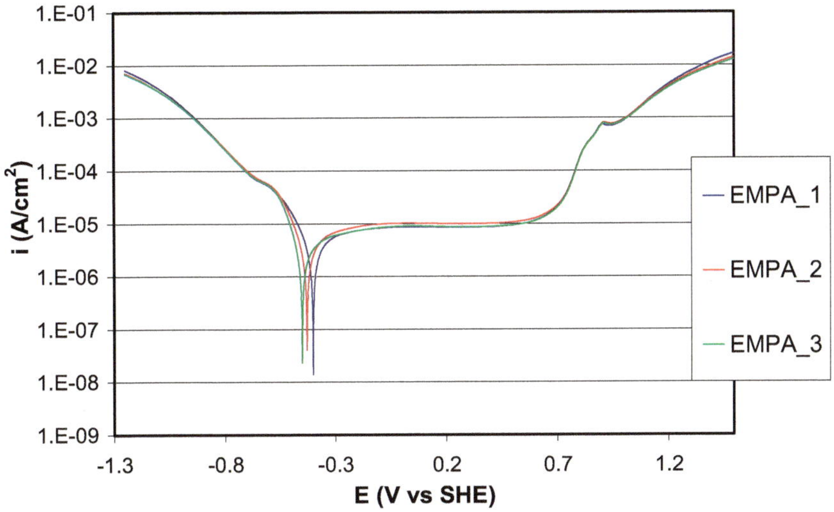
1.E-01
1.E-02
1.E-03
1.E-04
1.E-05
1.E-06
1.E-07
1.E-08
1.E-09
i (A/cm^2)
-1.3
-0.8
-0.3
0.2
0.7
1.2
E (V vs SHE)
EMPA_1
EMPA_2
EMPA_3

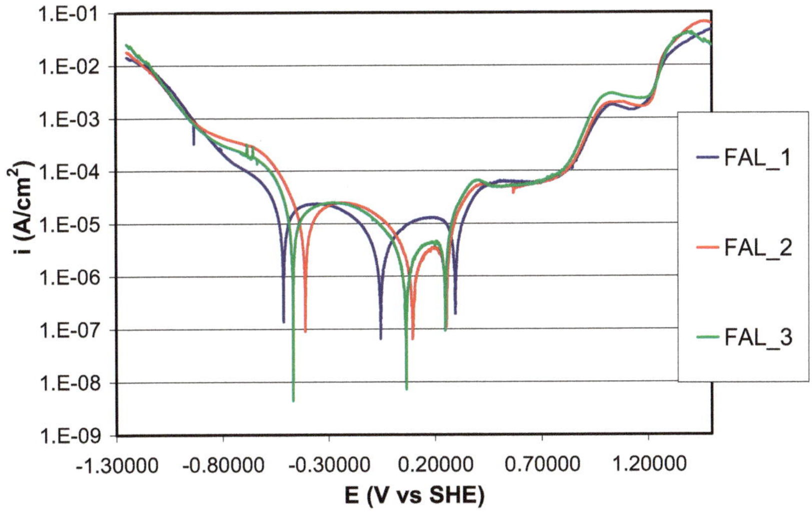
1.E-01
1.E-02
1.E-03
1.E-04
1.E-05
1.E-06
1.E-07
1.E-08
1.E-09
i (A/cm^2)
-1.30000
-0.80000
-0.30000
0.20000
0.70000
1.20000
E (V vs SHE)
FAL_1
FAL_2
FAL_3

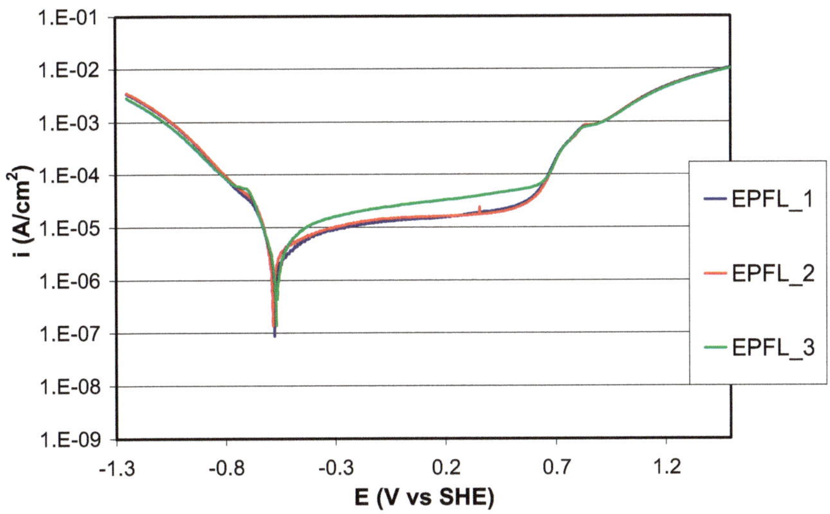
1.E-01
1.E-02
1.E-03
1.E-04
1.E-05
1.E-06
1.E-07
1.E-08
1.E-09
i (A/cm^2)
-1.3
-0.8
-0.3
0.2
0.7
1.2
E (V vs SHE)
EPFL_1
EPFL_2
EPFL_3

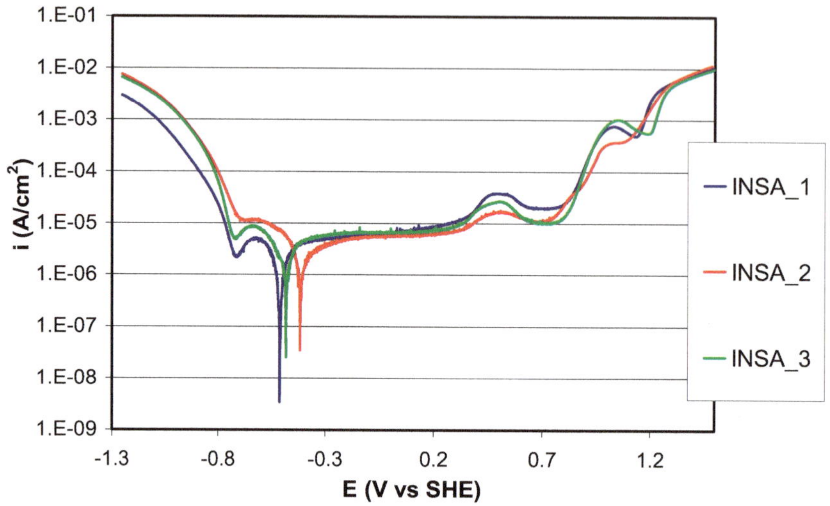
1.E-01
1.E-02
1.E-03
1.E-04
1.E-05
1.E-06
1.E-07
1.E-08
1.E-09
i (A/cm^2)
-1.3
-0.8
-0.3
0.2
0.7
1.2
E (V vs SHE)
INSA_1
INSA_2
INSA_3

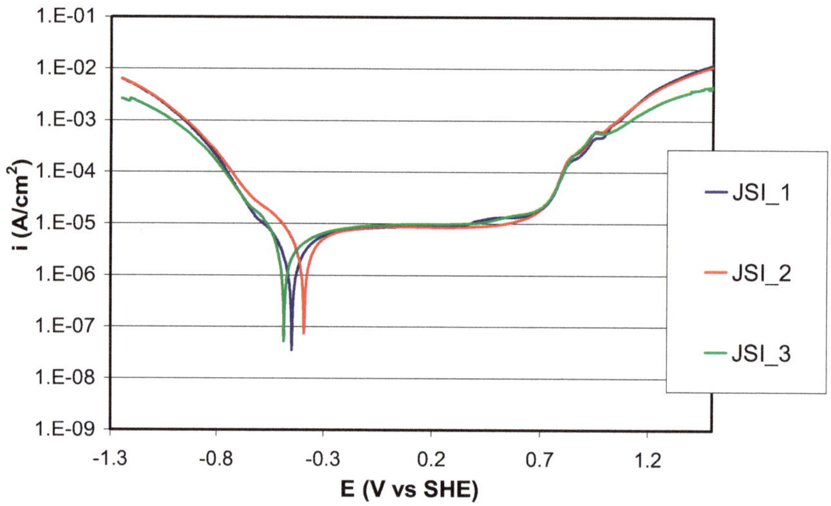
1.E-01
1.E-02
1.E-03
1.E-04
1.E-05
1.E-06
1.E-07
1.E-08
1.E-09
i (A/cm^2)
-1.3
-0.8
-0.3
0.2
0.7
1.2
E (V vs SHE)
JSI_1
JSI_2
JSI_3

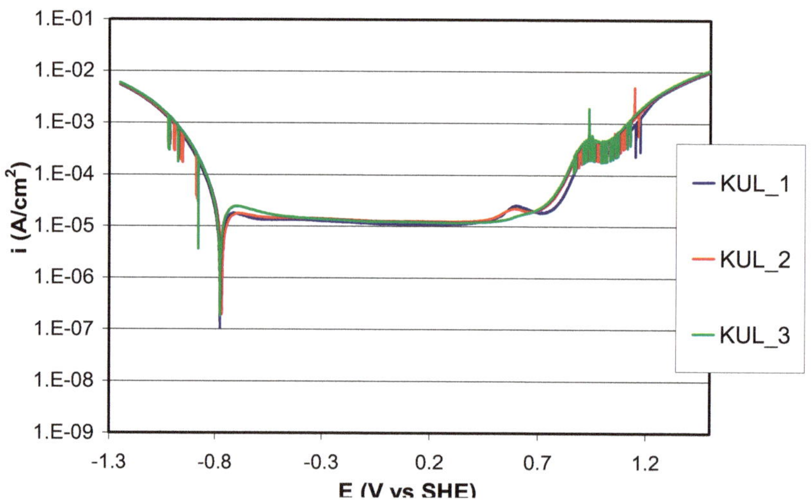
1.E-01
1.E-02
1.E-03
1.E-04
1.E-05
1.E-06
1.E-07
1.E-08
1.E-09
i (A/cm^2)
-1.3
-0.8
-0.3
0.2
0.7
1.2
KUL_1
KUL_2
KUL_3

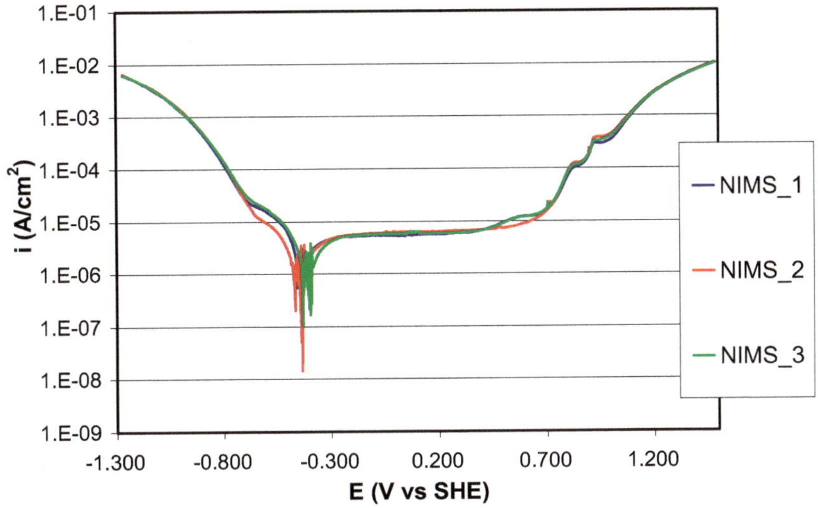
1.E-01
1.E-02
1.E-03
1.E-04
1.E-05
1.E-06
1.E-07
1.E-08
1.E-09
i (A/cm^2)
-1.300
-0.800
-0.300
0.200
0.700
1.200
E (V vs SHE)
NIMS_1
NIMS_2
NIMS_3

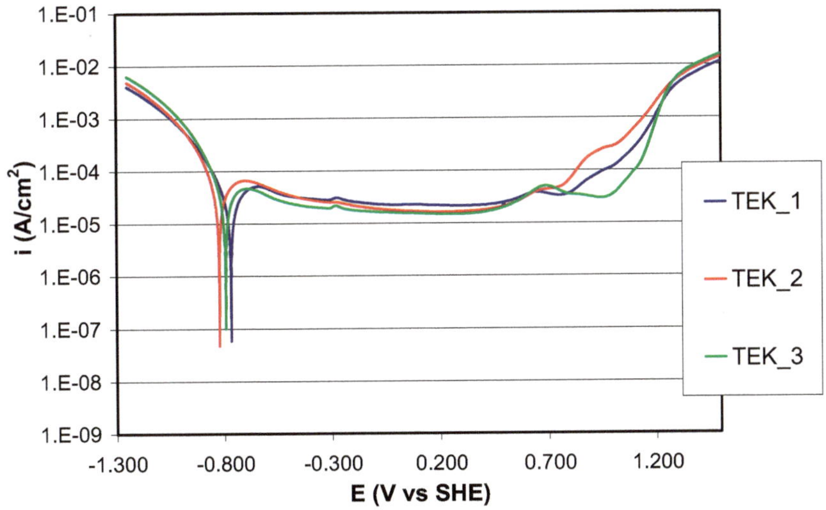
1.E-01
1.E-02
1.E-03
1.E-04
1.E-05
1.E-06
1.E-07
1.E-08
1.E-09
i (A/cm^2)
-1.300
-0.800
-0.300
0.200
0.700
1.200
E (V vs SHE)
TEK_1
TEK_2
TEK_3

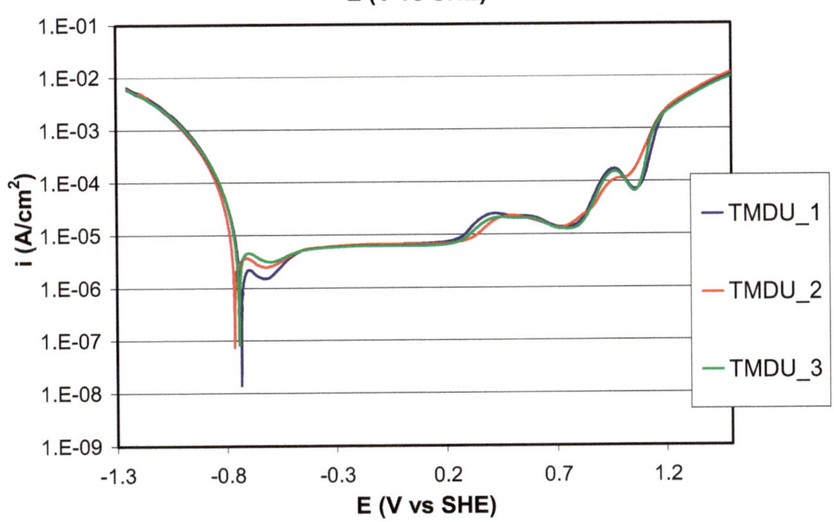
1.E-01
1.E-02
1.E-03
1.E-04
1.E-05
1.E-06
1.E-07
1.E-08
1.E-09
i (A/cm^2)
-1.3
-0.8
-0.3
0.2
0.7
1.2
E (V vs SHE)
TMDU_1
TMDU_2
TMDU_3

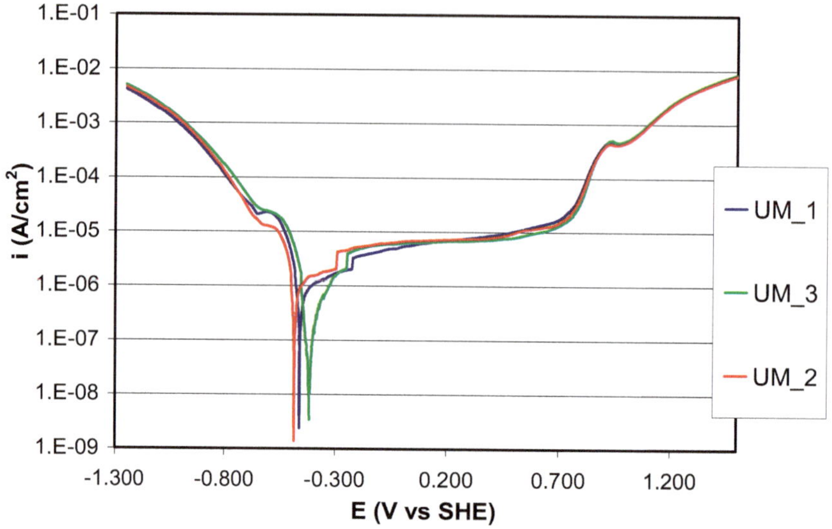
1.E-01
1.E-02
1.E-03
1.E-04
1.E-05
1.E-06
1.E-07
1.E-08
1.E-09
i (A/cm2)
-1.300
-0.800
-0.300
0.200
0.700
1.200
E (V vs SHE)
UM_1
UM_3
UM_2

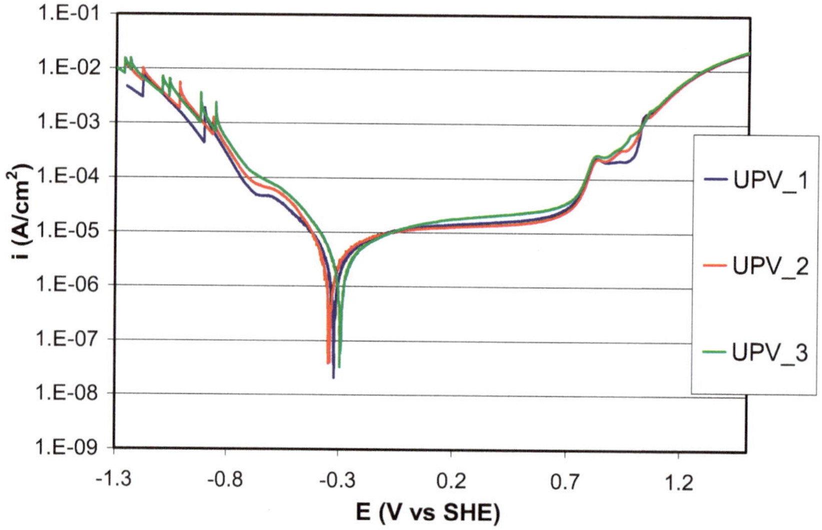
1.E-01
1.E-02
1.E-03
1.E-04
1.E-05
1.E-06
1.E-07
1.E-08
1.E-09
i (A/cm2)
-1.3
-0.8
-0.3
0.2
0.7
1.2
E (V vs SHE)
UPV_1
UPV_2
UPV_3

Appendix B
Alternating current (AC) results: Impedance spectra obtained by each participant laboratory at 0.15 V_{SHE} and OCP with and without albumin

B.1 Impedance data **at 0.15 V_{SHE}** in PBS **without albumin**: 3 repeated tests for each laboratory

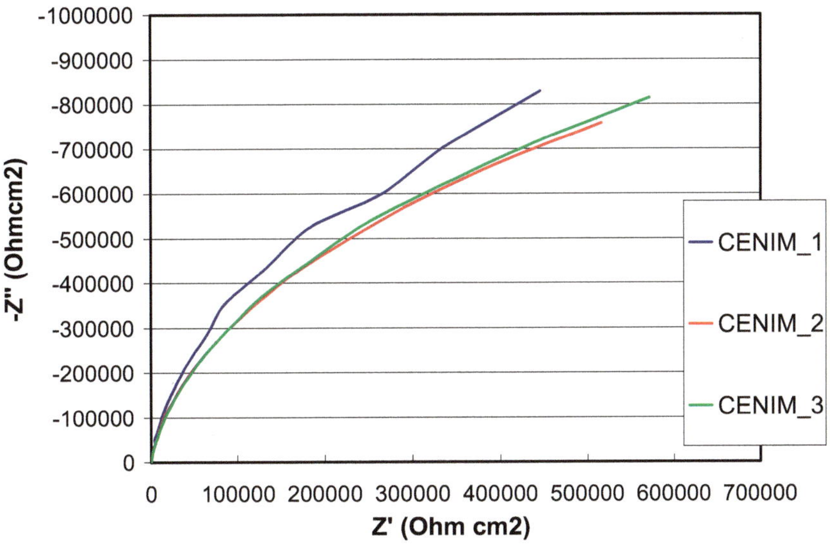

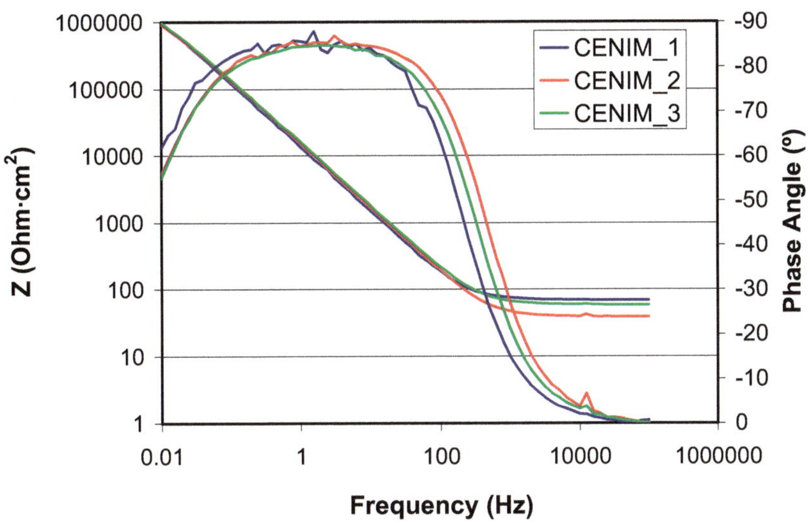

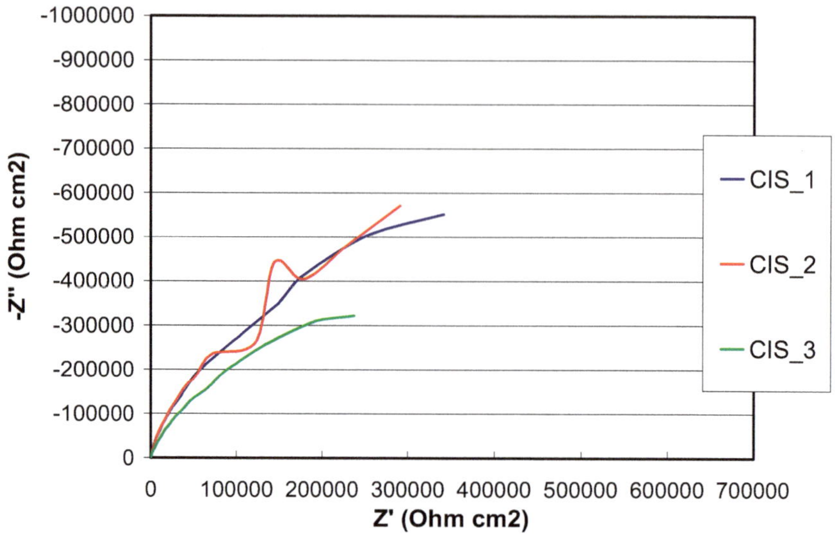
-Z'' (Ohm cm2)
Z' (Ohm cm2)
CIS_1
CIS_2
CIS_3

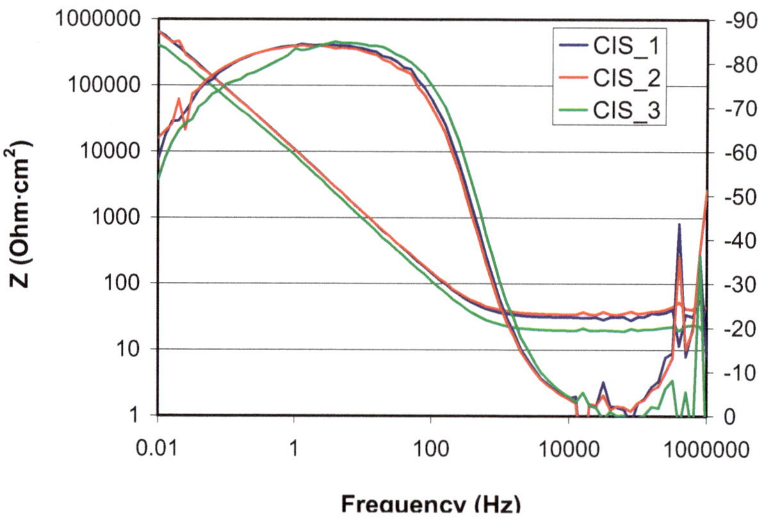
Z (Ohm·cm2)
Phase Angle (°)
Frequency (Hz)
CIS_1
CIS_2
CIS_3

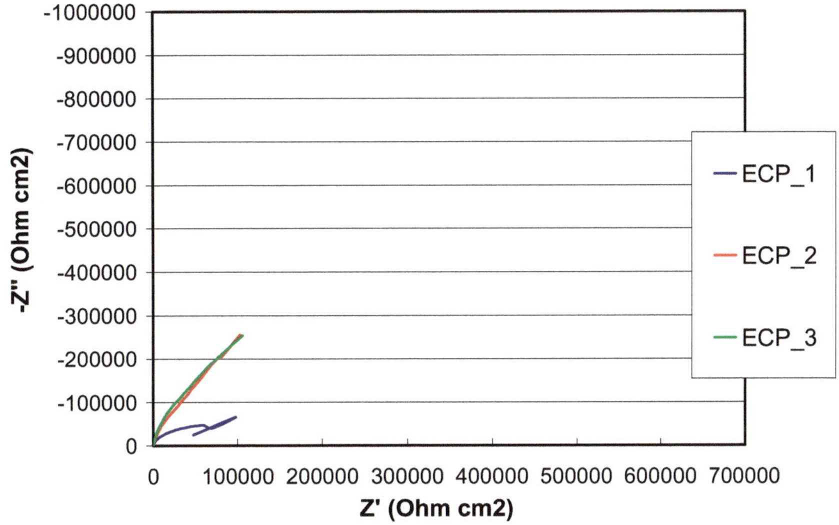
-1000000
-900000
-800000
-700000
-600000
-500000
-400000
-300000
-200000
-100000
0
-Z" (Ohm cm2)
0
100000
200000
300000
400000
500000
600000
700000
Z' (Ohm cm2)
ECP_1
ECP_2
ECP_3

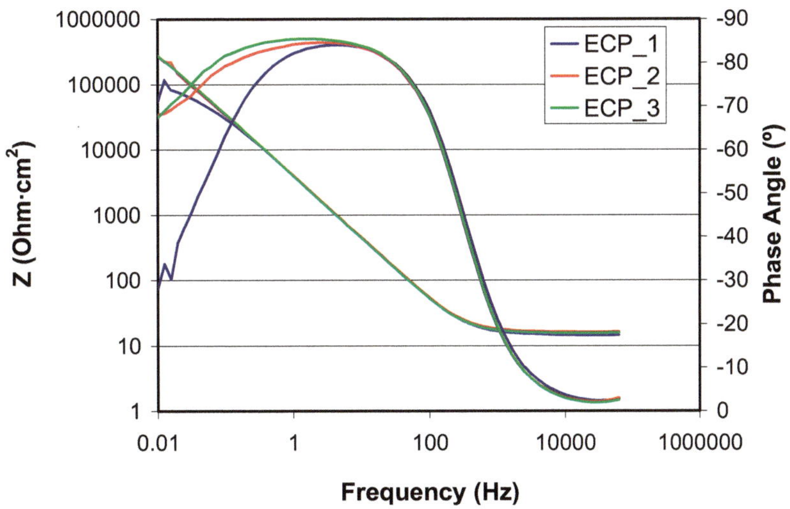
1000000
100000
10000
1000
100
10
1
Z (Ohm·cm^2)
0.01
1
100
10000
1000000
Frequency (Hz)
-90
-80
-70
-60
-50
-40
-30
-20
-10
0
Phase Angle (º)
ECP_1
ECP_2
ECP_3

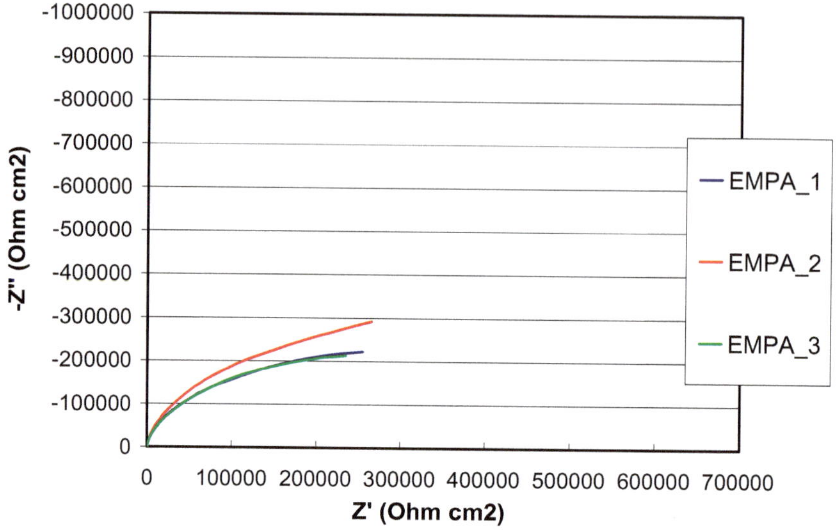
-1000000
-900000
-800000
-700000
-600000
-500000
-400000
-300000
-200000
-100000
0
-Z'' (Ohm cm2)
0
100000
200000
300000
400000
500000
600000
700000
Z' (Ohm cm2)
EMPA_1
EMPA_2
EMPA_3

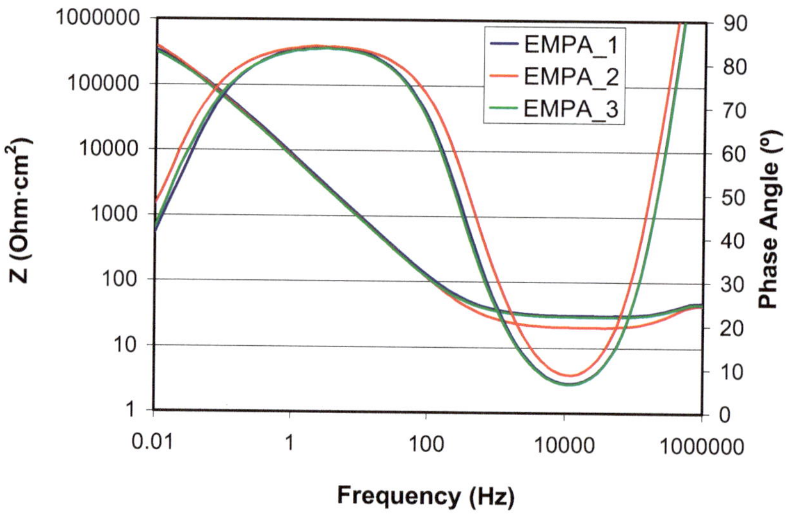
1000000
100000
10000
1000
100
10
1
Z (Ohm·cm^2)
0.01
1
100
10000
1000000
Frequency (Hz)
90
80
70
60
50
40
30
20
10
0
Phase Angle (°)
EMPA_1
EMPA_2
EMPA_3

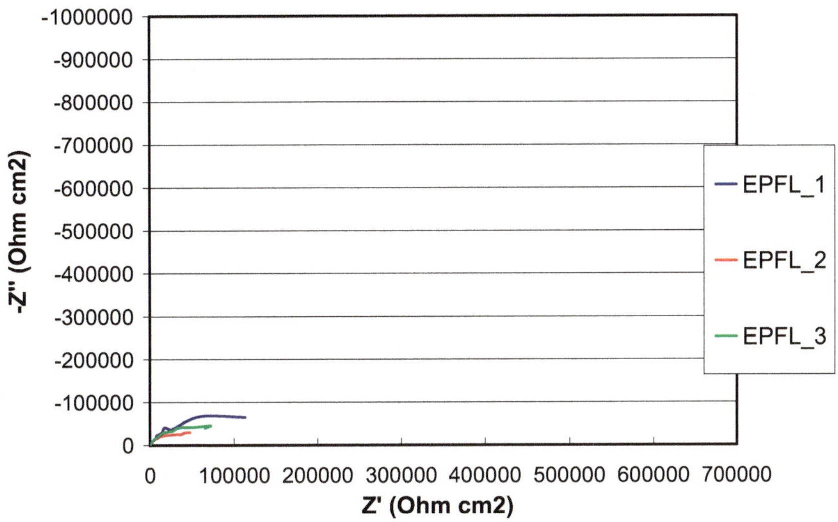
-1000000
-900000
-800000
-700000
-600000
-500000
-400000
-300000
-200000
-100000
0
-Z'' (Ohm cm2)
EPFL_1
EPFL_2
EPFL_3
0
100000
200000
300000
400000
500000
600000
700000
Z' (Ohm cm2)

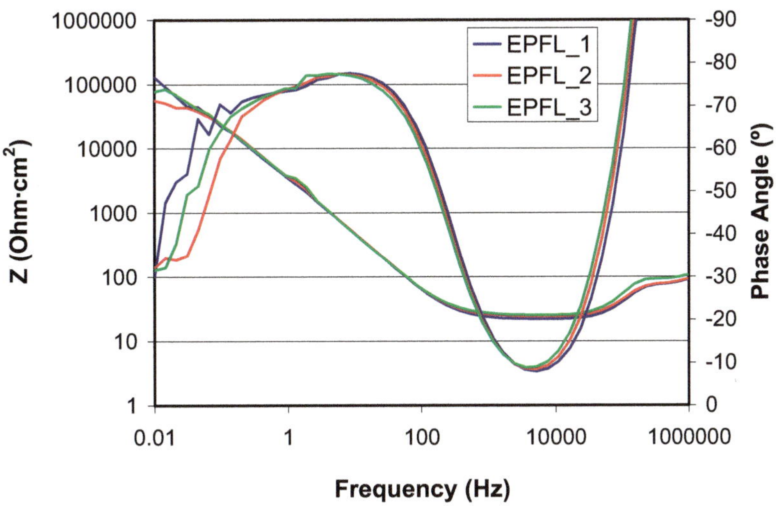
1000000
100000
10000
1000
100
10
1
Z (Ohm·cm2)
EPFL_1
EPFL_2
EPFL_3
-90
-80
-70
-60
-50
-40
-30
-20
-10
0
Phase Angle (°)
0.01
1
100
10000
1000000
Frequency (Hz)

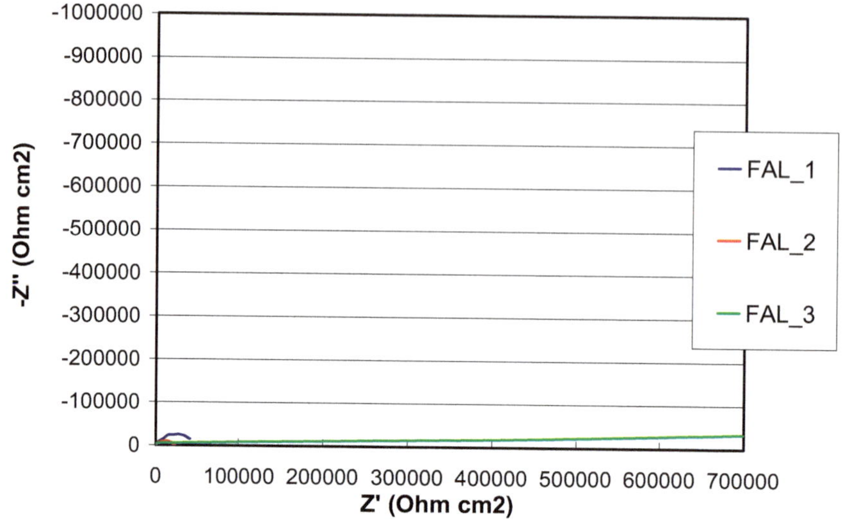
-Z'' (Ohm cm2)
-1000000
-900000
-800000
-700000
-600000
-500000
-400000
-300000
-200000
-100000
0
0
100000 200000 300000 400000 500000 600000 700000
Z' (Ohm cm2)
FAL_1
FAL_2
FAL_3

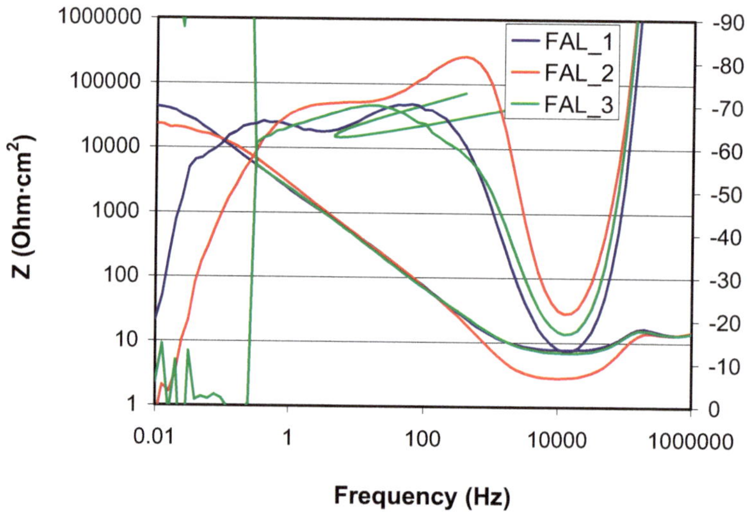
Z (Ohm·cm2)
1000000
100000
10000
1000
100
10
1
0.01
1
100
10000
1000000
Frequency (Hz)
Phase Angle (°)
-90
-80
-70
-60
-50
-40
-30
-20
-10
0
FAL_1
FAL_2
FAL_3

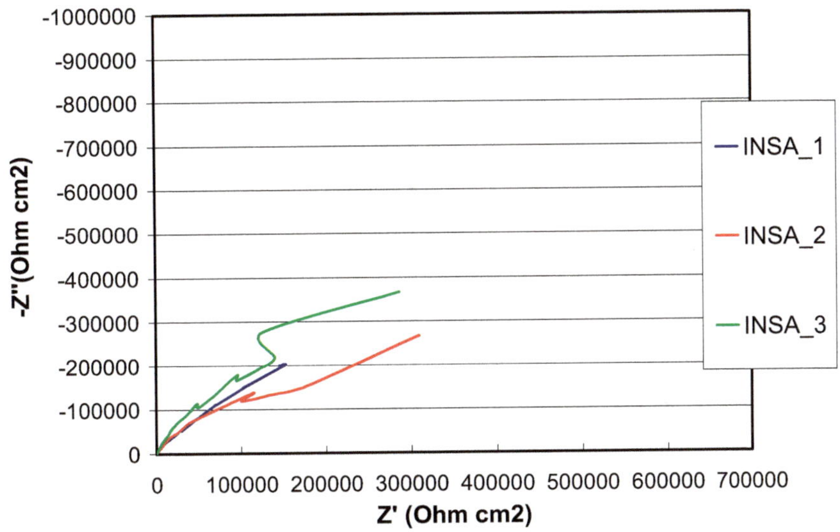
-1000000
-900000
-800000
-700000
-600000
-500000
-400000
-300000
-200000
-100000
0
-Z"(Ohm cm2)
0
100000
200000
300000
400000
500000
600000
700000
Z' (Ohm cm2)
INSA_1
INSA_2
INSA_3

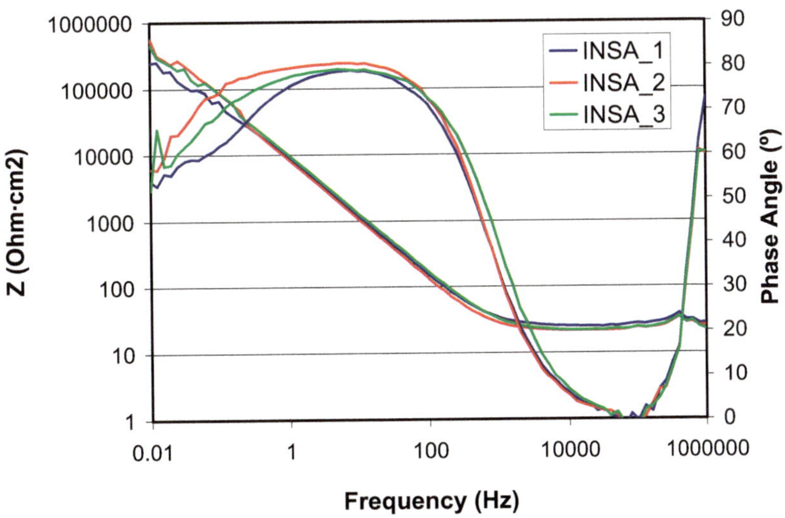
1000000
100000
10000
1000
100
10
1
Z (Ohm·cm2)
0.01
1
100
10000
1000000
Frequency (Hz)
90
80
70
60
50
40
30
20
10
0
Phase Angle (º)
INSA_1
INSA_2
INSA_3

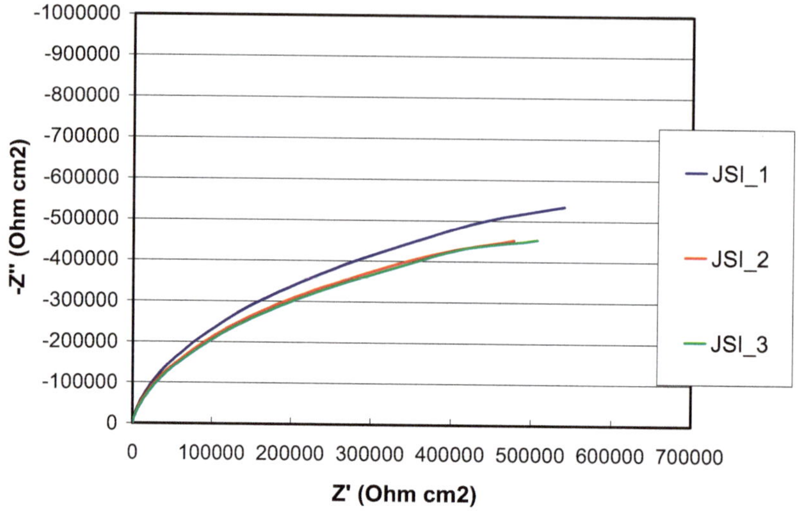
-Z'' (Ohm cm2)
-1000000
-900000
-800000
-700000
-600000
-500000
-400000
-300000
-200000
-100000
0
0
100000 200000 300000 400000 500000 600000 700000
Z' (Ohm cm2)
JSI_1
JSI_2
JSI_3

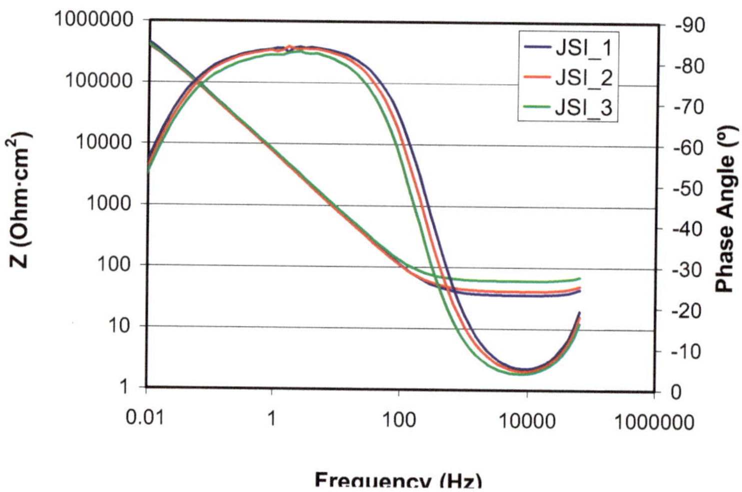
Z (Ohm·cm²)
1000000
100000
10000
1000
100
10
1
0.01
1
100
10000
1000000
Frequency (Hz)
Phase Angle (º)
-90
-80
-70
-60
-50
-40
-30
-20
-10
0
JSI_1
JSI_2
JSI_3

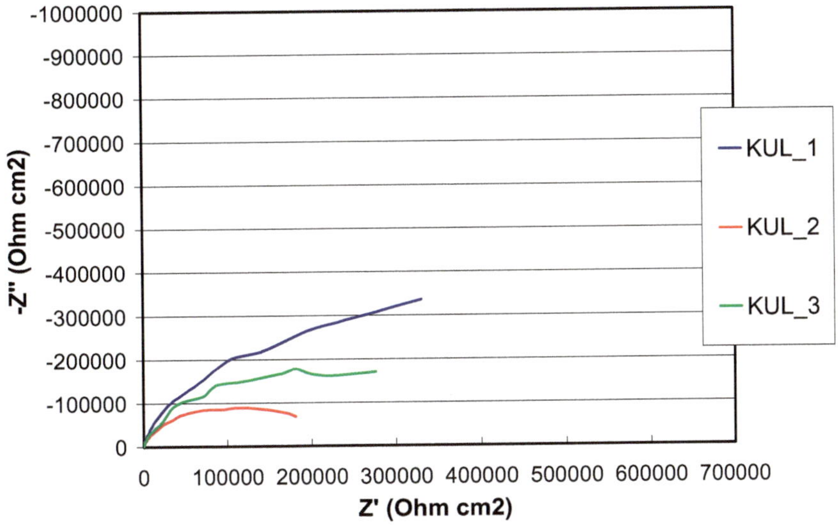
-1000000
-900000
-800000
-700000
-600000
-500000
-400000
-300000
-200000
-100000
0
-Z'' (Ohm cm2)
0
100000
200000
300000
400000
500000
600000
700000
Z' (Ohm cm2)
KUL_1
KUL_2
KUL_3

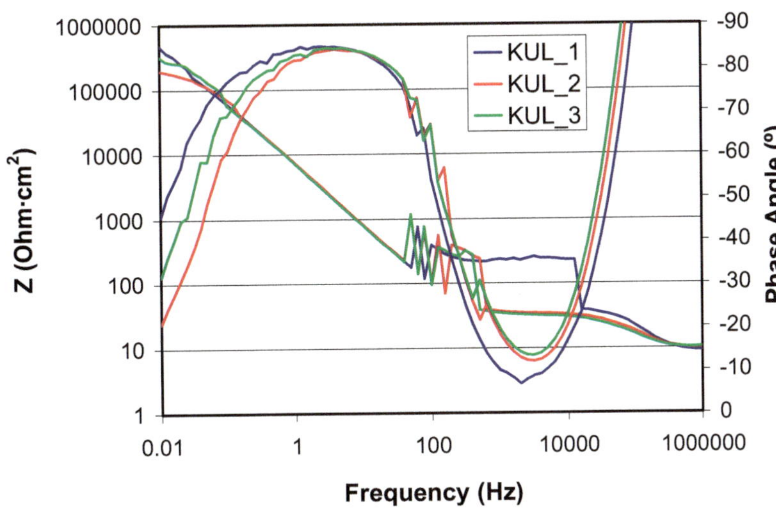
1000000
100000
10000
1000
100
10
1
Z (Ohm·cm2)
0.01
1
100
10000
1000000
Frequency (Hz)
-90
-80
-70
-60
-50
-40
-30
-20
-10
0
Phase Angle (°)
KUL_1
KUL_2
KUL_3

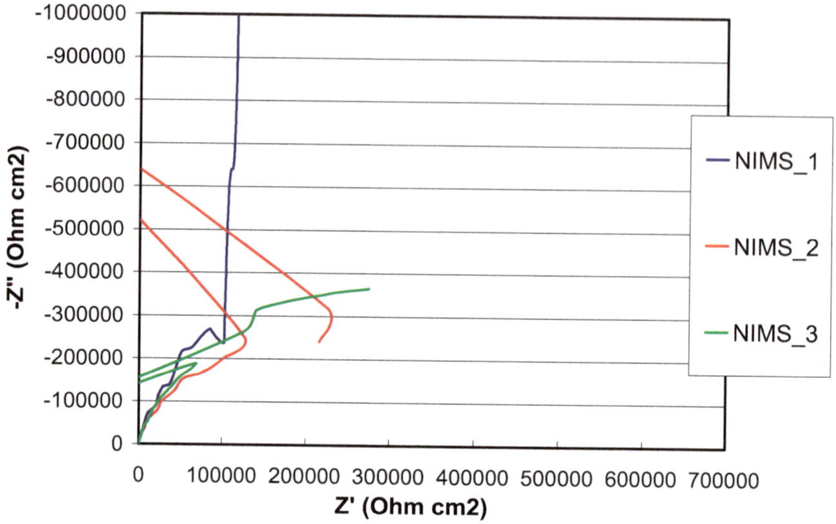
-Z" (Ohm cm2)
-1000000
-900000
-800000
-700000
-600000
-500000
-400000
-300000
-200000
-100000
0
0
100000
200000
300000
400000
500000
600000
700000
Z' (Ohm cm2)
NIMS_1
NIMS_2
NIMS_3

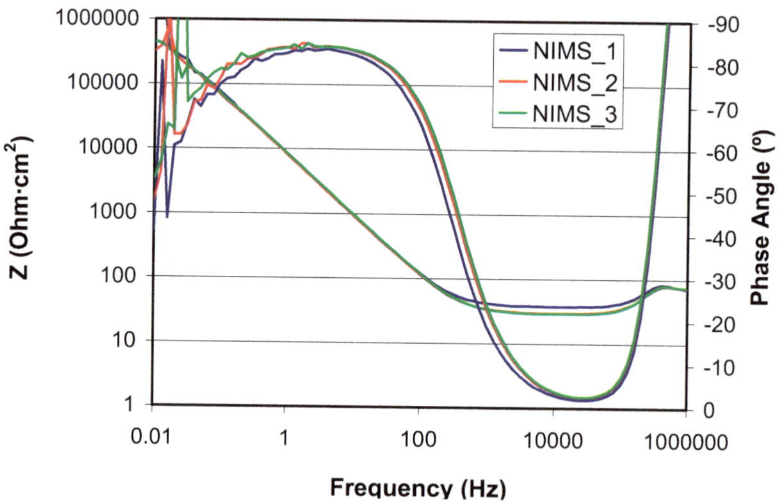
Z (Ohm·cm²)
1000000
100000
10000
1000
100
10
1
0.01
1
100
10000
1000000
Frequency (Hz)
Phase Angle (º)
-90
-80
-70
-60
-50
-40
-30
-20
-10
0
NIMS_1
NIMS_2
NIMS_3

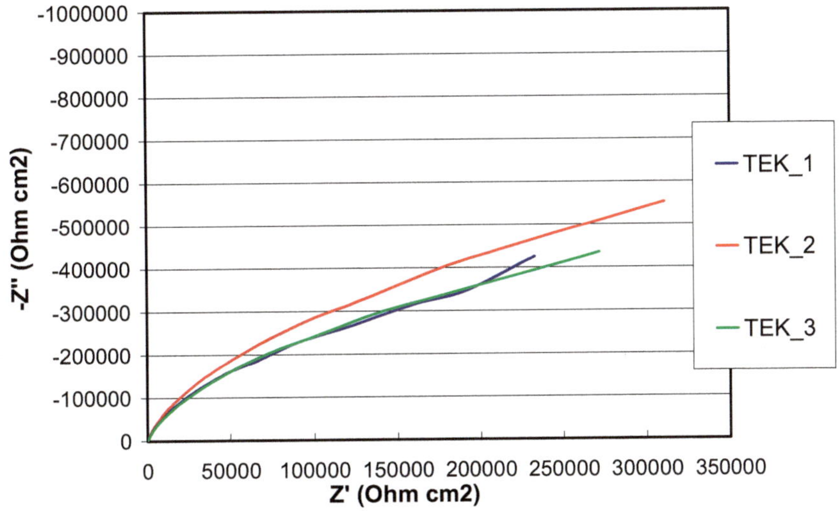
-1000000
-900000
-800000
-700000
-600000
-500000
-400000
-300000
-200000
-100000
0
-Z'' (Ohm cm2)
0
50000
100000
150000
200000
250000
300000
350000
Z' (Ohm cm2)
TEK_1
TEK_2
TEK_3

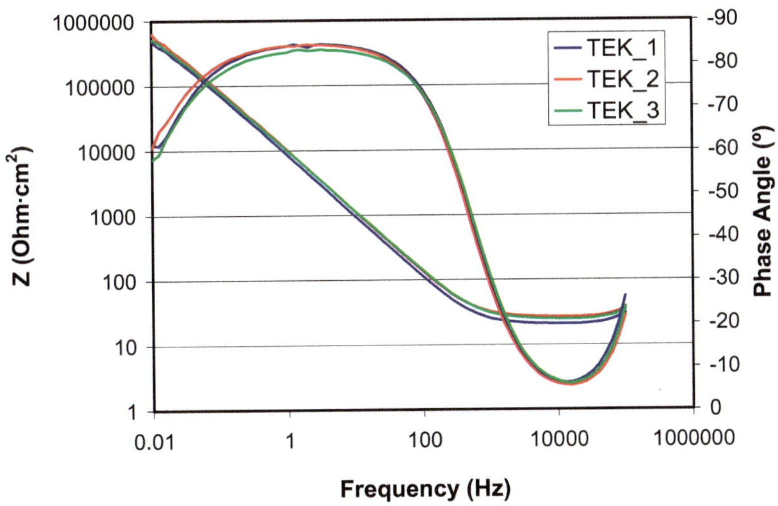
1000000
100000
10000
1000
100
10
1
Z (Ohm·cm2)
0.01
1
100
10000
1000000
Frequency (Hz)
-90
-80
-70
-60
-50
-40
-30
-20
-10
0
Phase Angle (°)
TEK_1
TEK_2
TEK_3

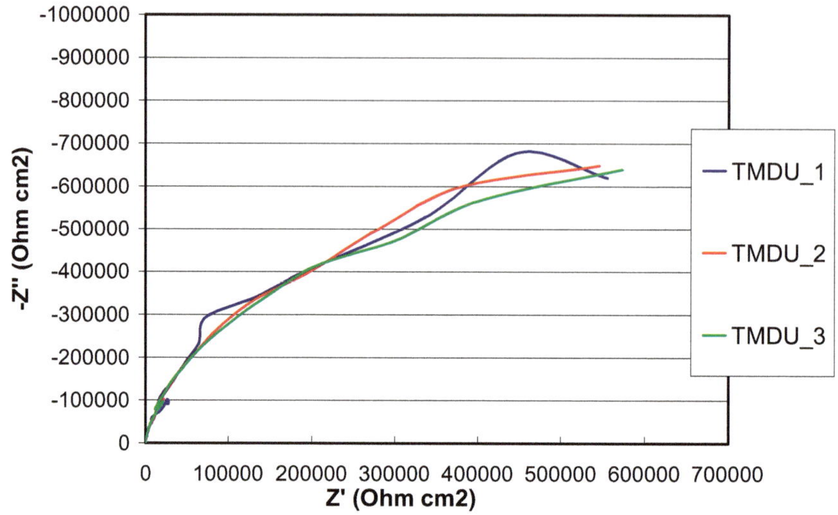
-1000000
-900000
-800000
-700000
-600000
-500000
-400000
-300000
-200000
-100000
0
-Z'' (Ohm cm2)
0
100000
200000
300000
400000
500000
600000
700000
Z' (Ohm cm2)
TMDU_1
TMDU_2
TMDU_3

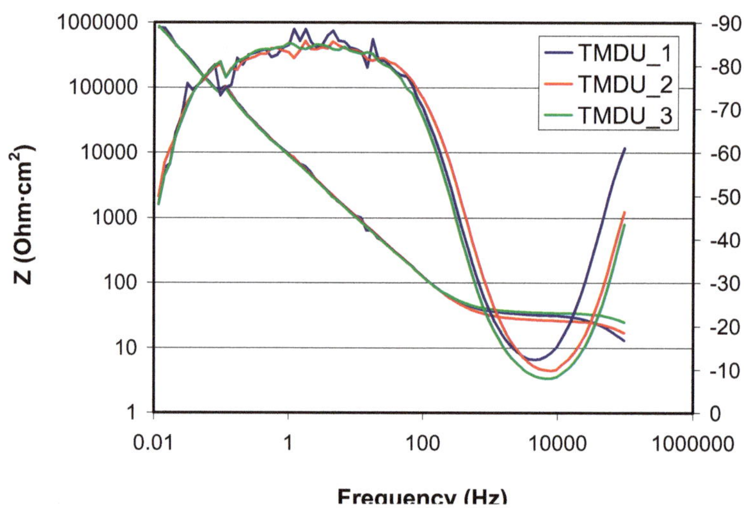
1000000
100000
10000
1000
100
10
1
Z (Ohm·cm^2)
-90
-80
-70
-60
-50
-40
-30
-20
-10
0
Phase Angle (º)
0.01
1
100
10000
1000000
Frequency (Hz)
TMDU_1
TMDU_2
TMDU_3

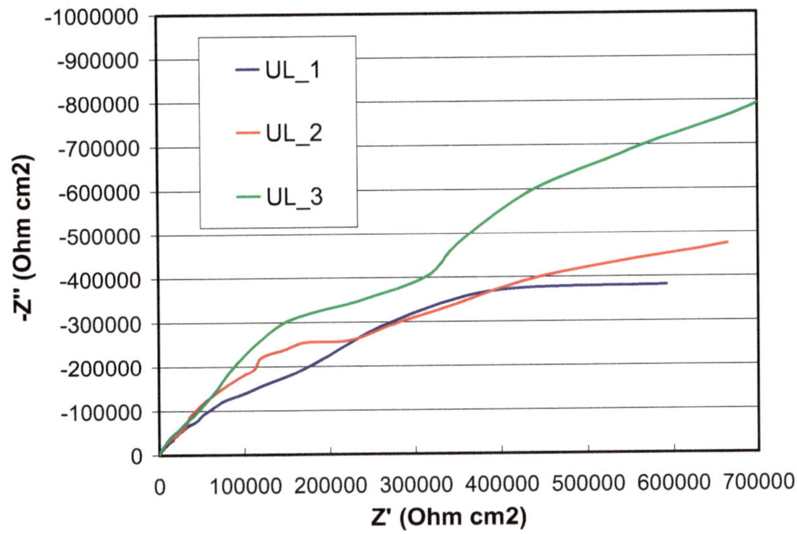
-1000000
-900000
-800000
-700000
-600000
-500000
-400000
-300000
-200000
-100000
0
-Z'' (Ohm cm2)
UL_1
UL_2
UL_3
0
100000
200000
300000
400000
500000
600000
700000
Z' (Ohm cm2)

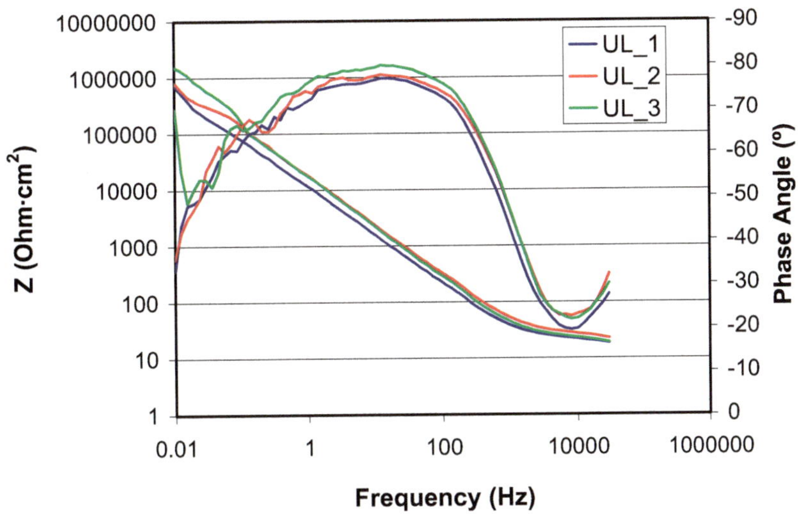
10000000
1000000
100000
10000
1000
100
10
1
Z (Ohm·cm2)
-90
-80
-70
-60
-50
-40
-30
-20
-10
0
Phase Angle (°)
UL_1
UL_2
UL_3
0.01
1
100
10000
1000000
Frequency (Hz)

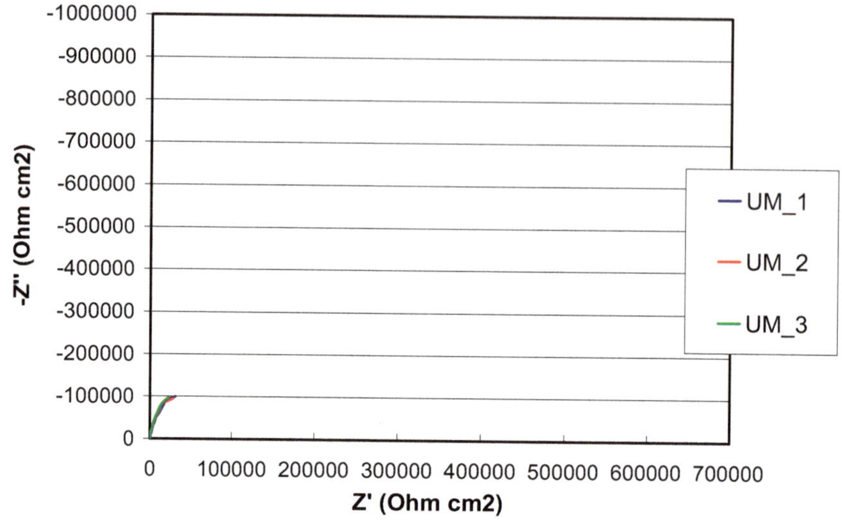
-Z'' (Ohm cm2)
Z' (Ohm cm2)
-1000000
-900000
-800000
-700000
-600000
-500000
-400000
-300000
-200000
-100000
0
100000 200000 300000 400000 500000 600000 700000
UM_1
UM_2
UM_3

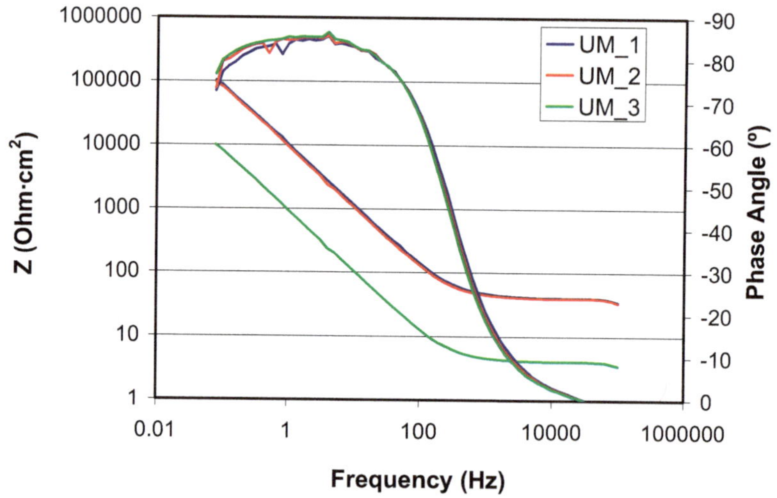
Z (Ohm·cm^2)
Phase Angle (°)
Frequency (Hz)
1000000
100000
10000
1000
100
10
1
0.01
1
100
10000
1000000
-90
-80
-70
-60
-50
-40
-30
-20
-10
0
UM_1
UM_2
UM_3

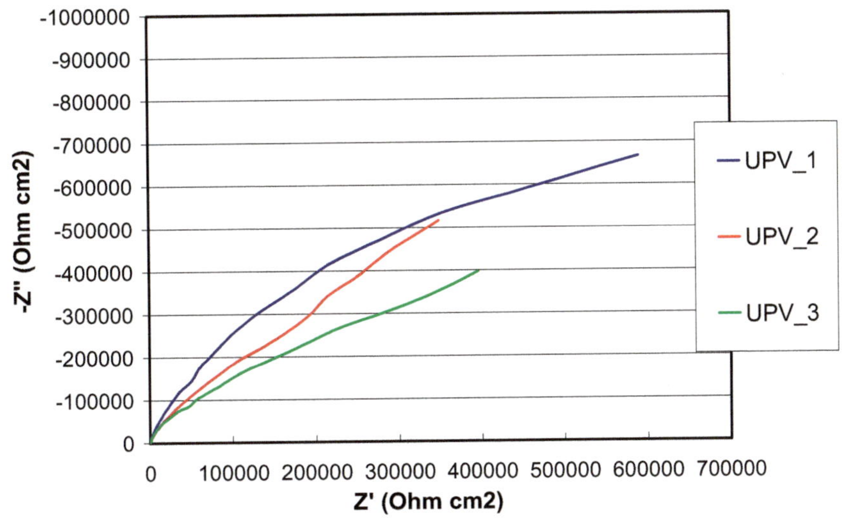
-Z'' (Ohm cm2)
Z' (Ohm cm2)
UPV_1
UPV_2
UPV_3
-1000000
-900000
-800000
-700000
-600000
-500000
-400000
-300000
-200000
-100000
0
100000 200000 300000 400000 500000 600000 700000

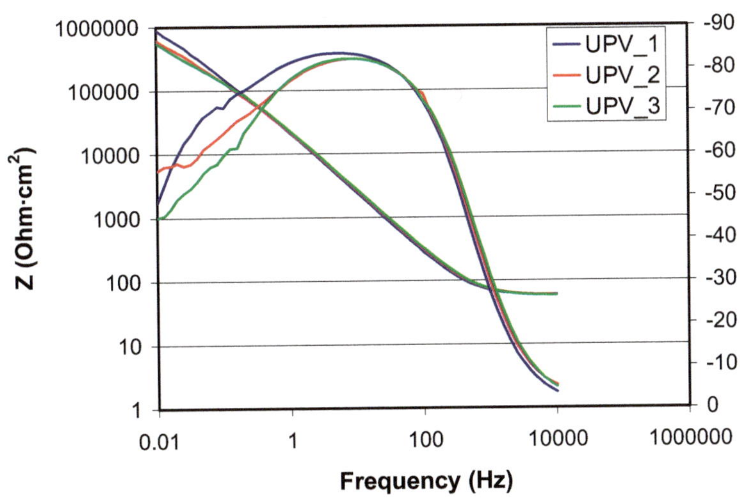
Z (Ohm·cm²)
Phase Angle (°)
Frequency (Hz)
UPV_1
UPV_2
UPV_3
1000000
100000
10000
1000
100
10
1
0.01
1
100
10000
1000000
-90
-80
-70
-60
-50
-40
-30
-20
-10
0

B.2 Impedance data **at OCP** in PBS **without albumin**: 3 repeated tests for each laboratory

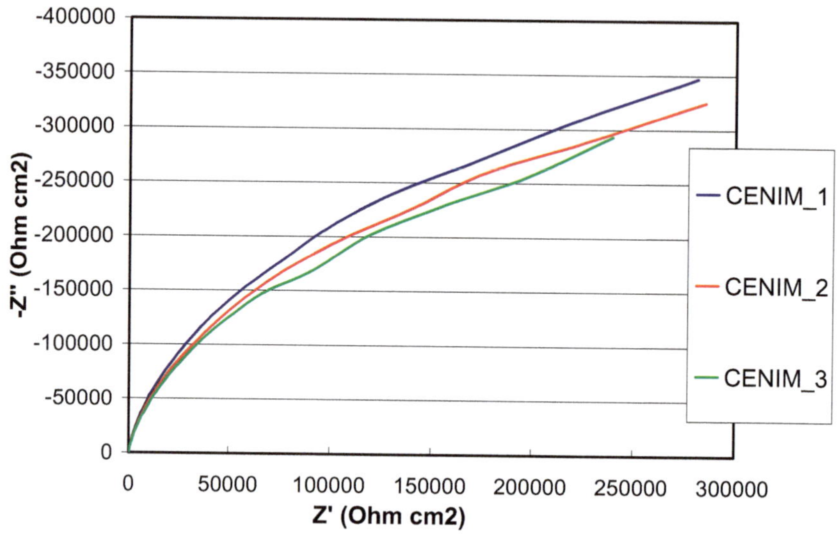

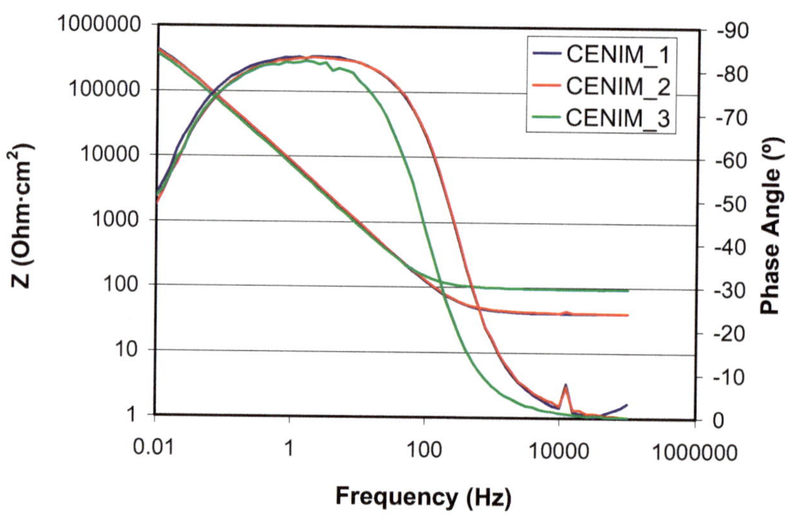

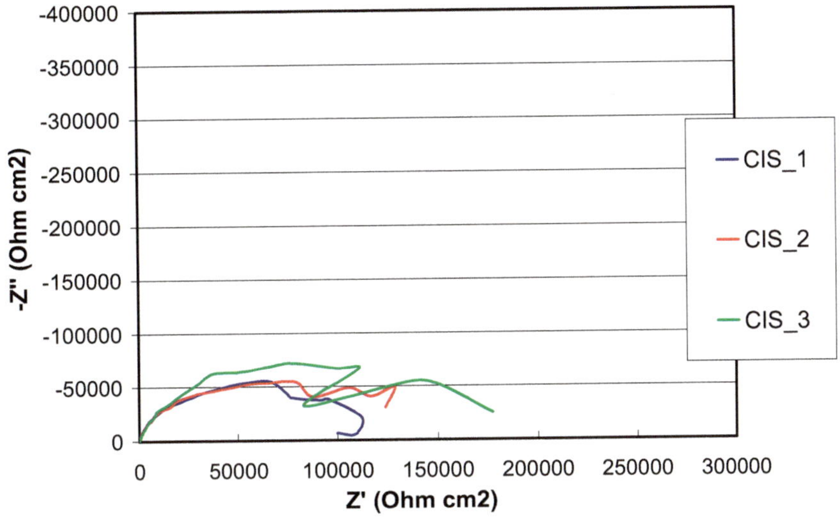
-400000
-350000
-300000
-250000
-200000
-150000
-100000
-50000
0
-Z'' (Ohm cm2)
0
50000
100000
150000
200000
250000
300000
Z' (Ohm cm2)
CIS_1
CIS_2
CIS_3

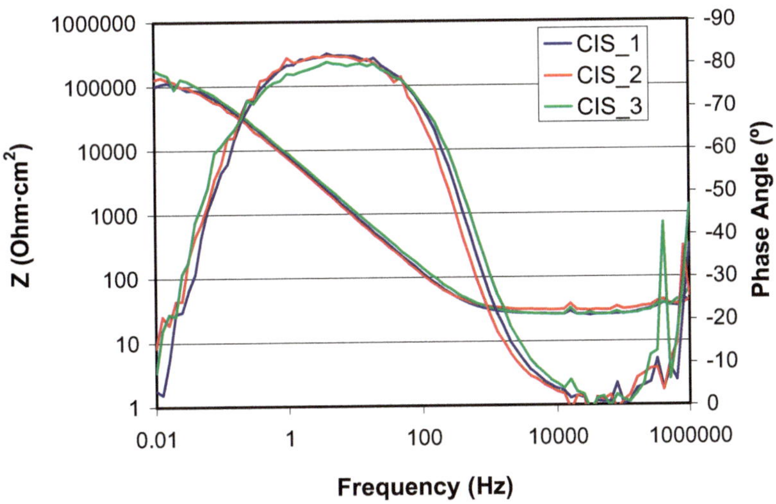
1000000
100000
10000
1000
100
10
1
Z (Ohm·cm²)
0.01
1
100
10000
1000000
Frequency (Hz)
-90
-80
-70
-60
-50
-40
-30
-20
-10
0
Phase Angle (°)
CIS_1
CIS_2
CIS_3

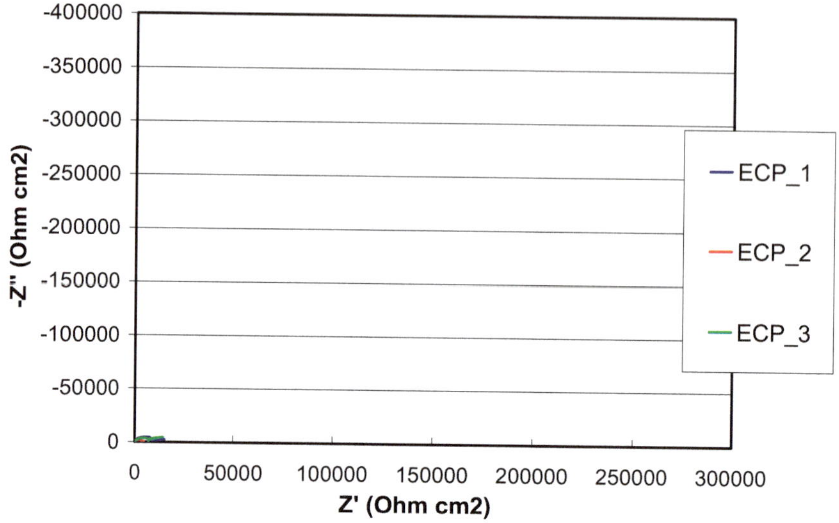
-400000
-350000
-300000
-250000
-200000
-150000
-100000
-50000
0
-Z'' (Ohm cm2)
0
50000
100000
150000
200000
250000
300000
Z' (Ohm cm2)
ECP_1
ECP_2
ECP_3

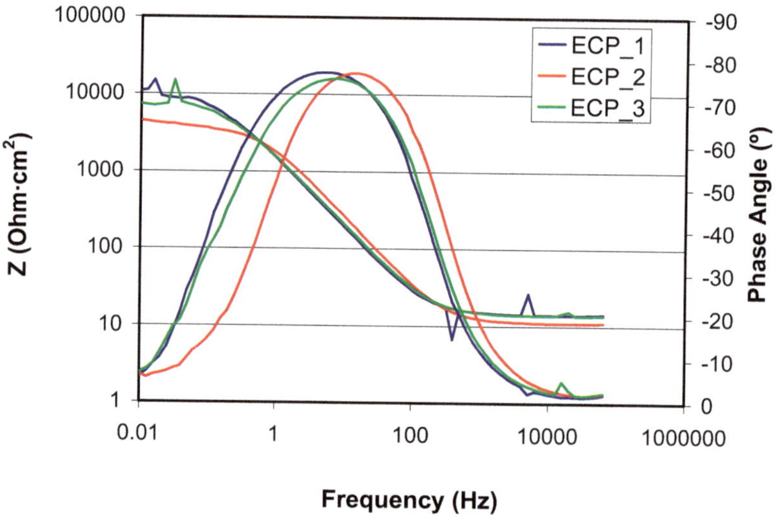
100000
10000
1000
100
10
1
Z (Ohm·cm2)
0.01
1
100
10000
1000000
Frequency (Hz)
-90
-80
-70
-60
-50
-40
-30
-20
-10
0
Phase Angle (º)
ECP_1
ECP_2
ECP_3

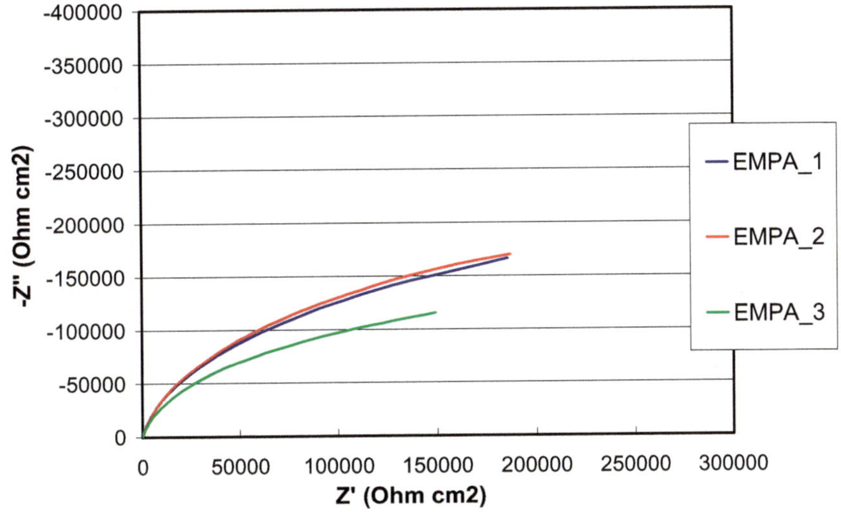
-400000
-350000
-300000
-250000
-200000
-150000
-100000
-50000
0
-Z'' (Ohm cm2)
0
50000
100000
150000
200000
250000
300000
Z' (Ohm cm2)
EMPA_1
EMPA_2
EMPA_3

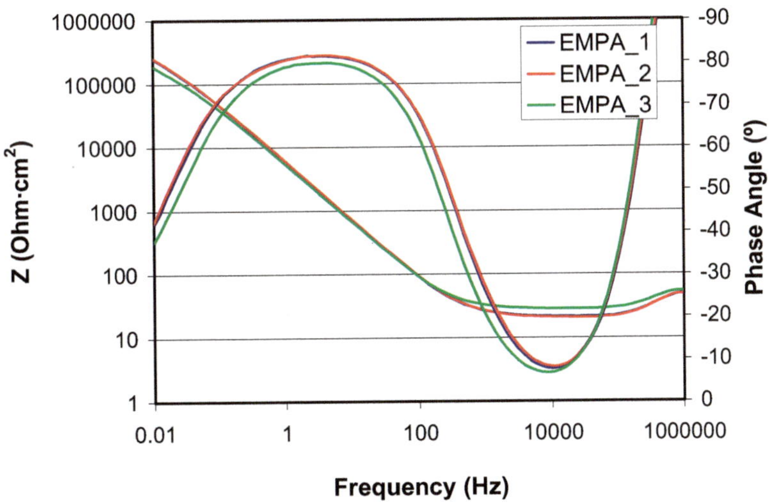
1000000
100000
10000
1000
100
10
1
Z (Ohm·cm^2)
0.01
1
100
10000
1000000
Frequency (Hz)
-90
-80
-70
-60
-50
-40
-30
-20
-10
0
Phase Angle (º)
EMPA_1
EMPA_2
EMPA_3

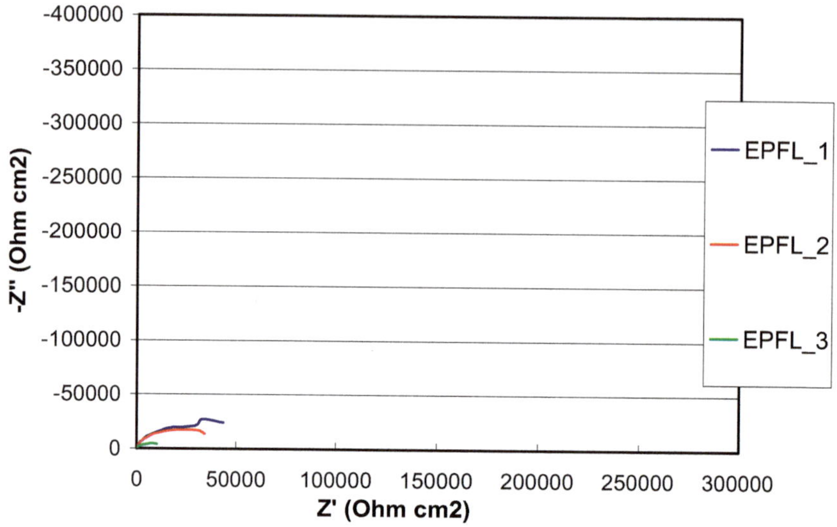
-400000
-350000
-300000
-250000
-200000
-150000
-100000
-50000
0
-Z'' (Ohm cm2)
0
50000
100000
150000
200000
250000
300000
Z' (Ohm cm2)
EPFL_1
EPFL_2
EPFL_3

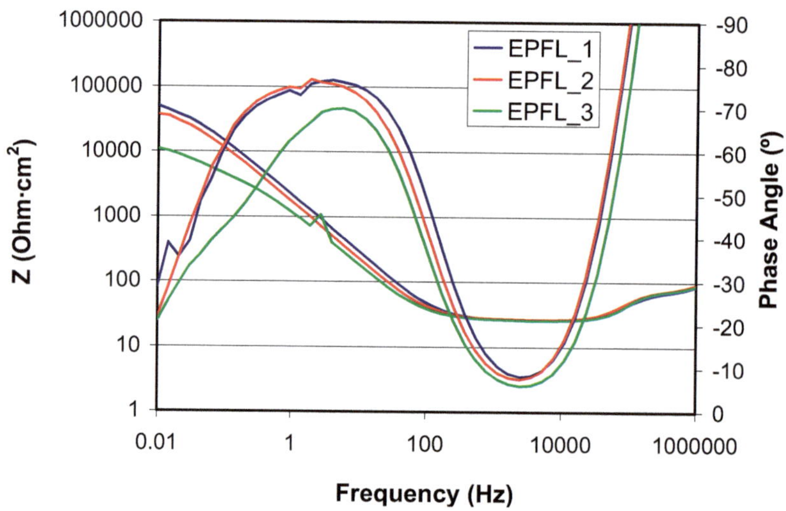
1000000
100000
10000
1000
100
10
1
Z (Ohm·cm2)
0.01
1
100
10000
1000000
Frequency (Hz)
-90
-80
-70
-60
-50
-40
-30
-20
-10
0
Phase Angle (º)
EPFL_1
EPFL_2
EPFL_3

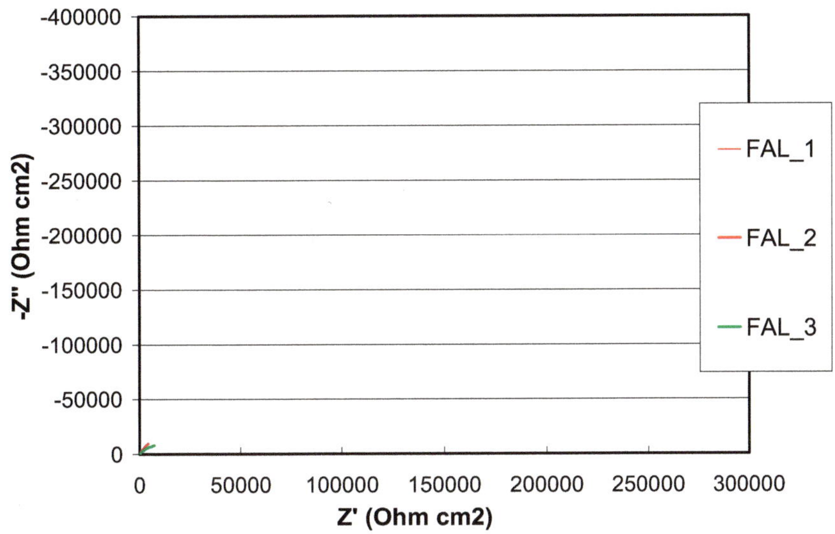
-400000
-350000
-300000
-250000
-200000
-150000
-100000
-50000
0
-Z'' (Ohm cm2)
0
50000
100000
150000
200000
250000
300000
Z' (Ohm cm2)
FAL_1
FAL_2
FAL_3

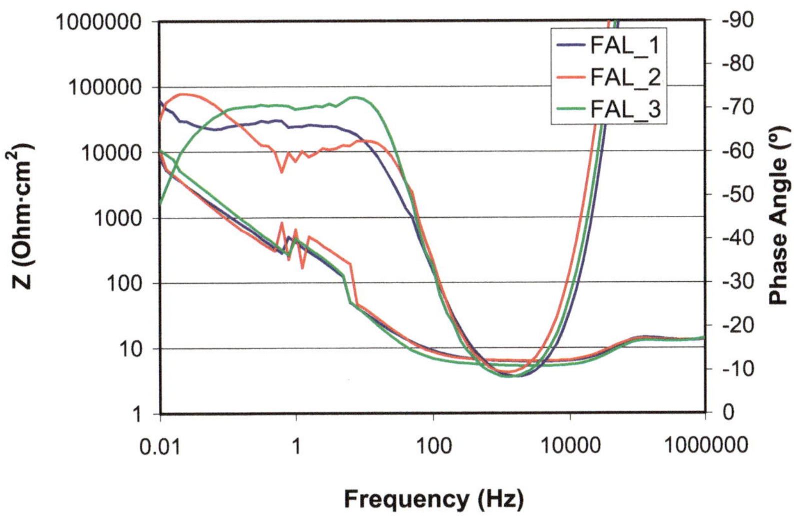
1000000
100000
10000
1000
100
10
1
Z (Ohm·cm²)
0.01
1
100
10000
1000000
Frequency (Hz)
-90
-80
-70
-60
-50
-40
-30
-20
-10
0
Phase Angle (°)
FAL_1
FAL_2
FAL_3

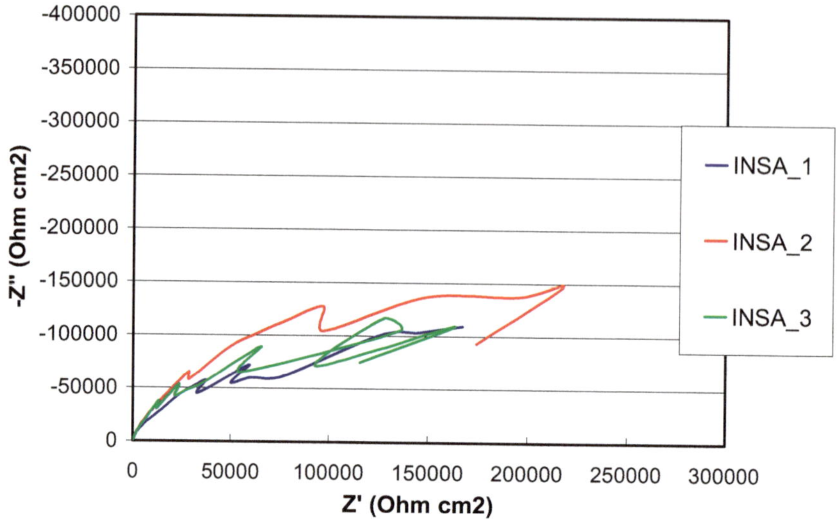
-400000
-350000
-300000
-250000
-200000
-150000
-100000
-50000
0
-Z'' (Ohm cm2)
0
50000
100000
150000
200000
250000
300000
Z' (Ohm cm2)
INSA_1
INSA_2
INSA_3

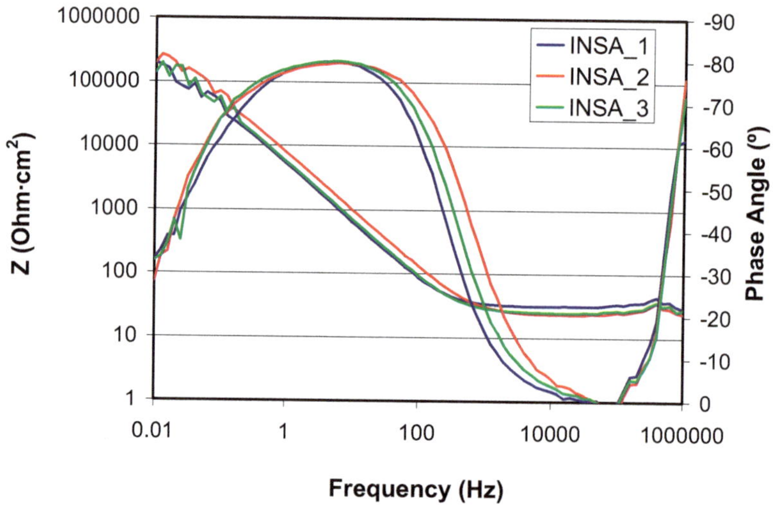
1000000
100000
10000
1000
100
10
1
Z (Ohm·cm²)
0.01
1
100
10000
1000000
Frequency (Hz)
-90
-80
-70
-60
-50
-40
-30
-20
-10
0
Phase Angle (º)
INSA_1
INSA_2
INSA_3

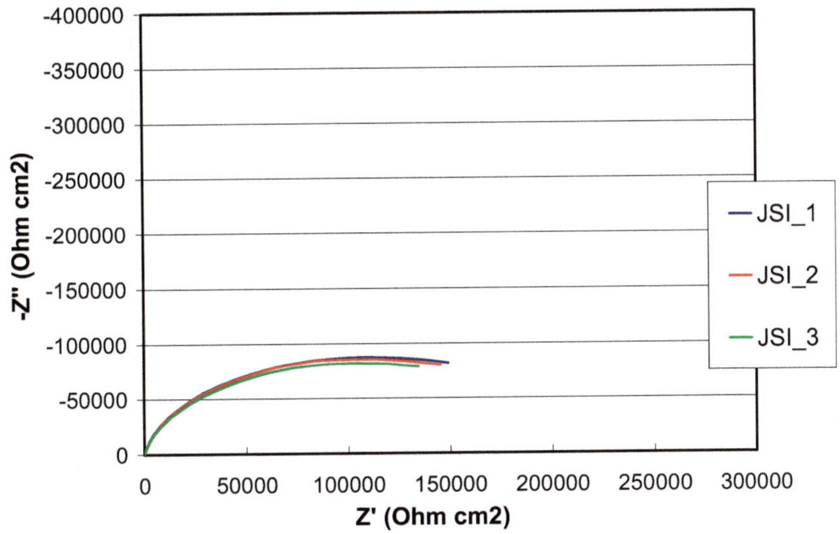
-400000
-350000
-300000
-250000
-200000
-150000
-100000
-50000
0
-Z" (Ohm cm2)
0
50000
100000
150000
200000
250000
300000
Z' (Ohm cm2)
JSI_1
JSI_2
JSI_3

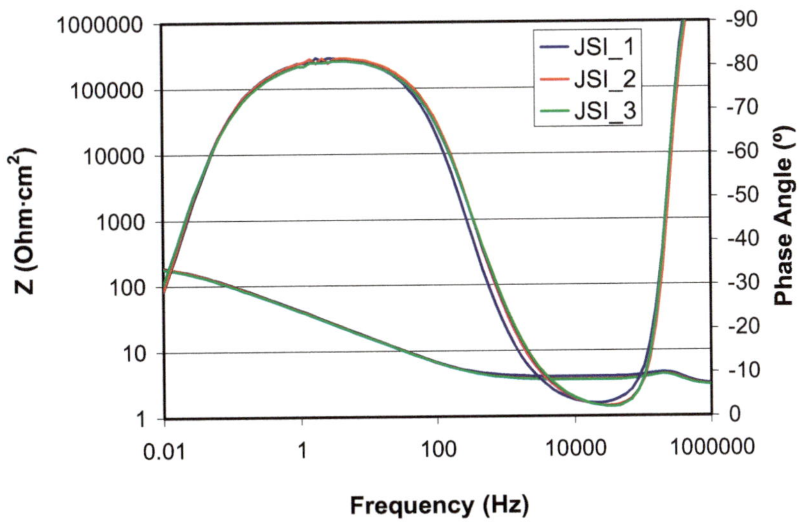
1000000
100000
10000
1000
100
10
1
Z (Ohm·cm2)
0.01
1
100
10000
1000000
Frequency (Hz)
-90
-80
-70
-60
-50
-40
-30
-20
-10
0
Phase Angle (°)
JSI_1
JSI_2
JSI_3

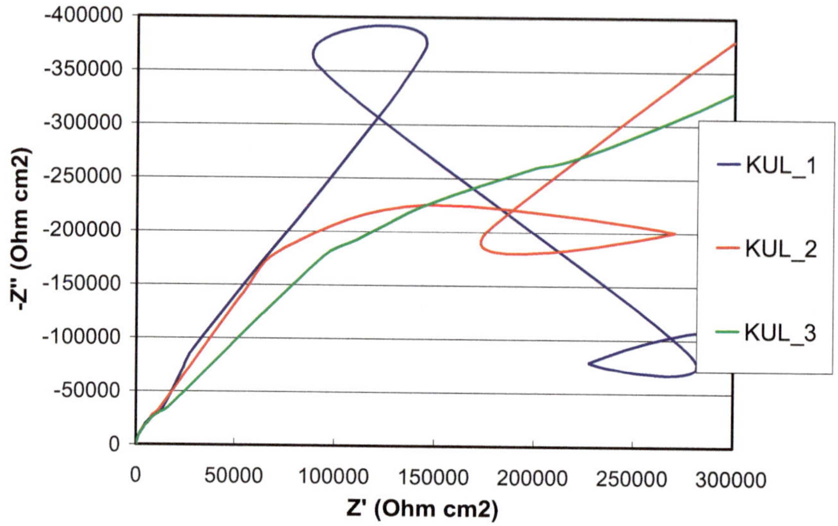
-400000
-350000
-300000
-250000
-200000
-150000
-100000
-50000
0
-Z'' (Ohm cm2)
0
50000
100000
150000
200000
250000
300000
Z' (Ohm cm2)
KUL_1
KUL_2
KUL_3

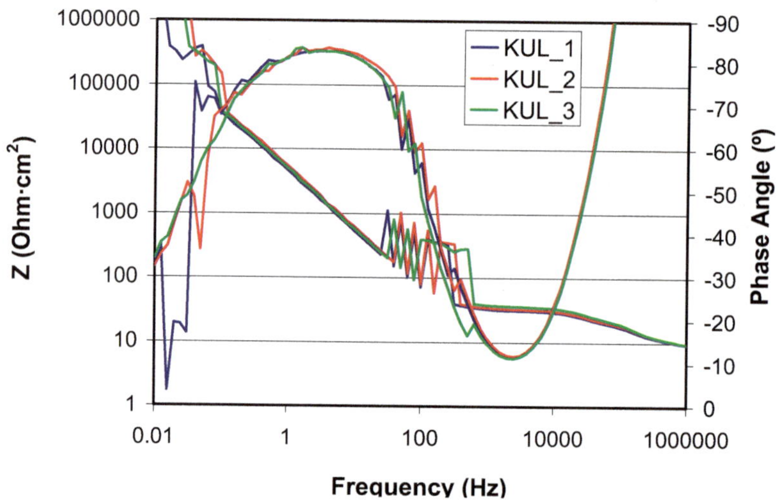
1000000
100000
10000
1000
100
10
1
Z (Ohm·cm2)
0.01
1
100
10000
1000000
Frequency (Hz)
-90
-80
-70
-60
-50
-40
-30
-20
-10
0
Phase Angle (º)
KUL_1
KUL_2
KUL_3

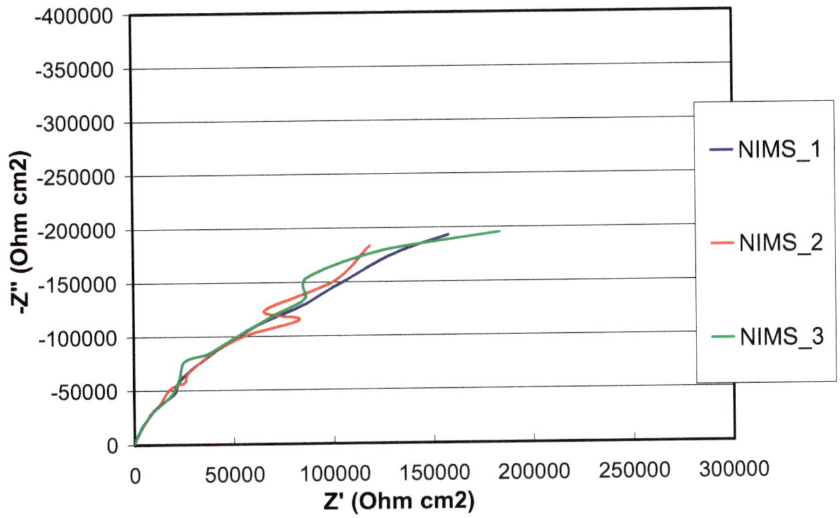
-400000
-350000
-300000
-250000
-200000
-150000
-100000
-50000
0
-Z'' (Ohm cm2)
0
50000
100000
150000
200000
250000
300000
Z' (Ohm cm2)
NIMS_1
NIMS_2
NIMS_3

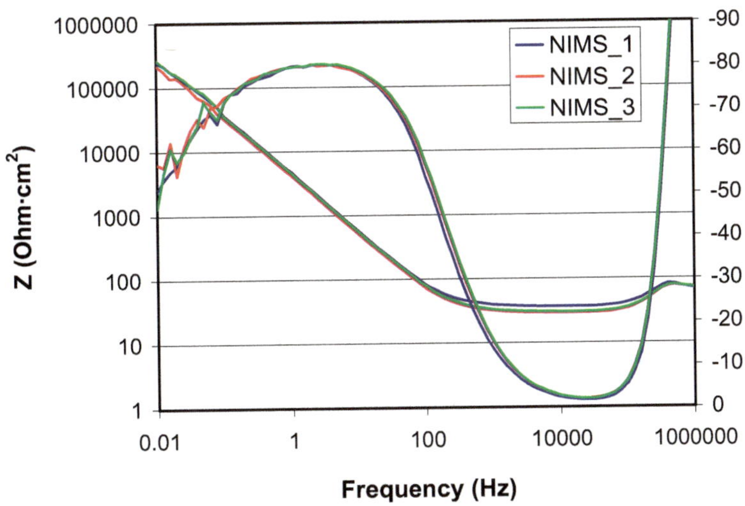
1000000
100000
10000
1000
100
10
1
Z (Ohm·cm2)
0.01
1
100
10000
1000000
Frequency (Hz)
-90
-80
-70
-60
-50
-40
-30
-20
-10
0
Phase Angle (°)
NIMS_1
NIMS_2
NIMS_3

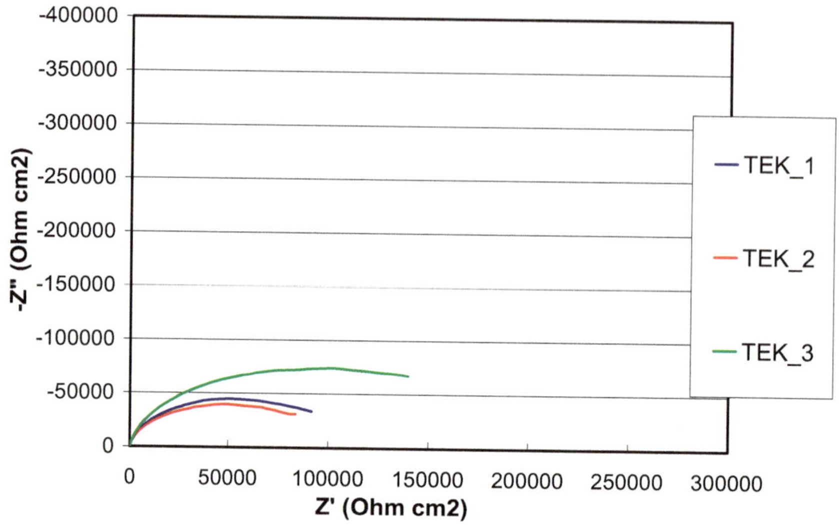
-400000
-350000
-300000
-250000
-200000
-150000
-100000
-50000
0
-Z" (Ohm cm2)
0
50000
100000
150000
200000
250000
300000
Z' (Ohm cm2)
TEK_1
TEK_2
TEK_3

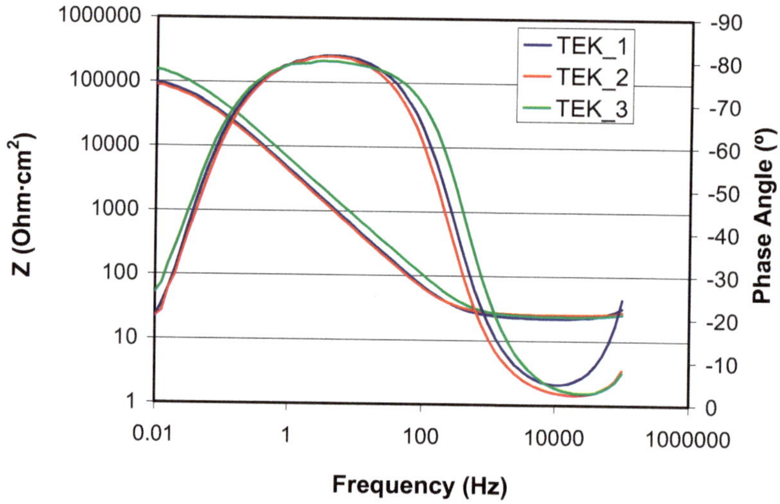
1000000
100000
10000
1000
100
10
1
Z (Ohm·cm²)
0.01
1
100
10000
1000000
Frequency (Hz)
-90
-80
-70
-60
-50
-40
-30
-20
-10
0
Phase Angle (°)
TEK_1
TEK_2
TEK_3

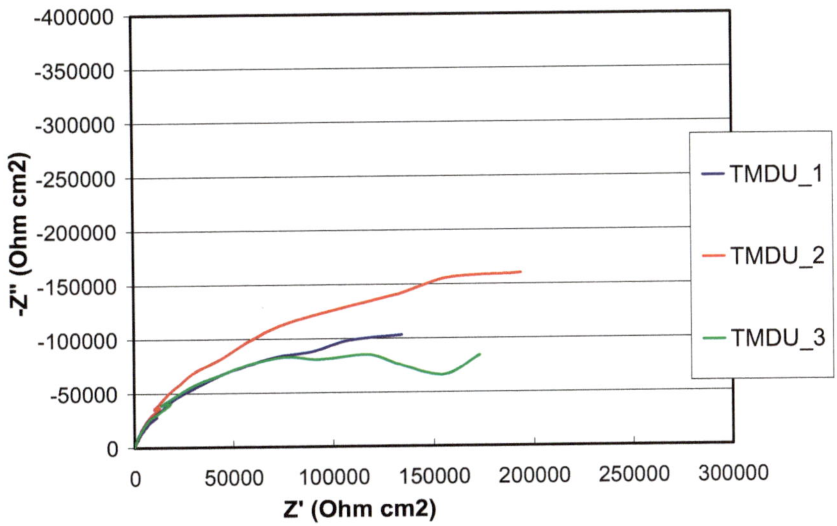
-400000
-350000
-300000
-250000
-200000
-150000
-100000
-50000
0
-Z" (Ohm cm2)
0
50000
100000
150000
200000
250000
300000
Z' (Ohm cm2)
TMDU_1
TMDU_2
TMDU_3

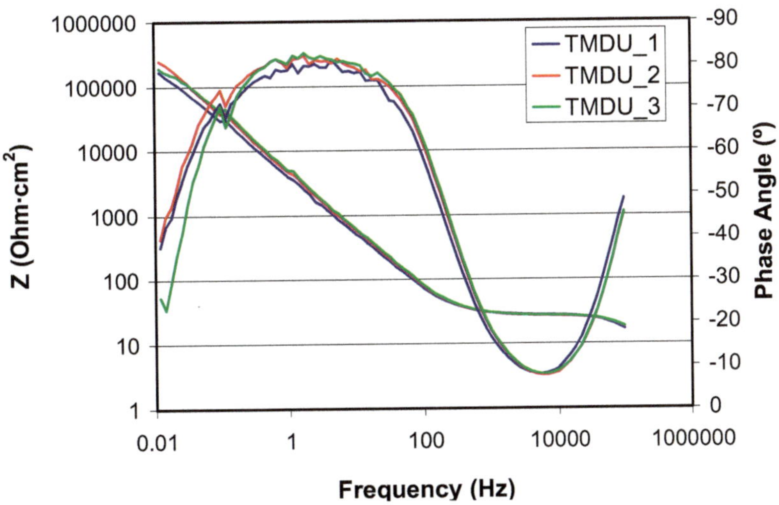
1000000
100000
10000
1000
100
10
1
Z (Ohm·cm2)
0.01
1
100
10000
1000000
Frequency (Hz)
-90
-80
-70
-60
-50
-40
-30
-20
-10
0
Phase Angle (º)
TMDU_1
TMDU_2
TMDU_3

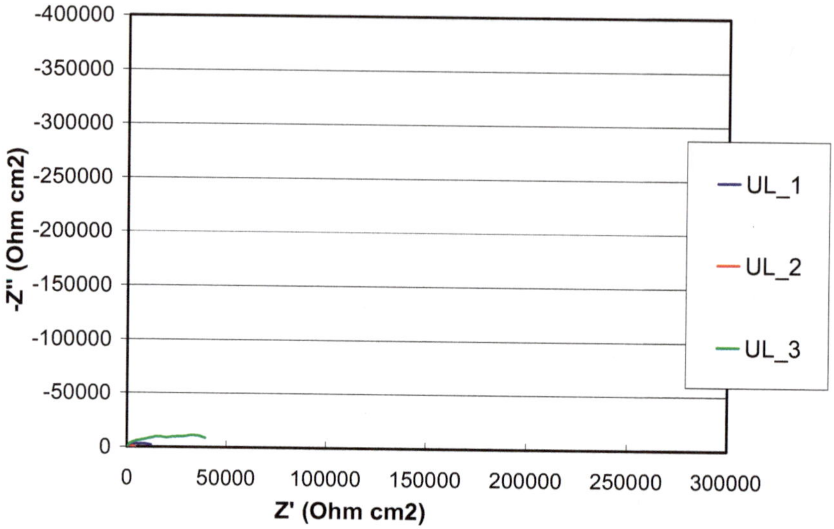
-400000
-350000
-300000
-250000
-200000
-150000
-100000
-50000
0
-Z'' (Ohm cm2)
0
50000
100000
150000
200000
250000
300000
Z' (Ohm cm2)
UL_1
UL_2
UL_3

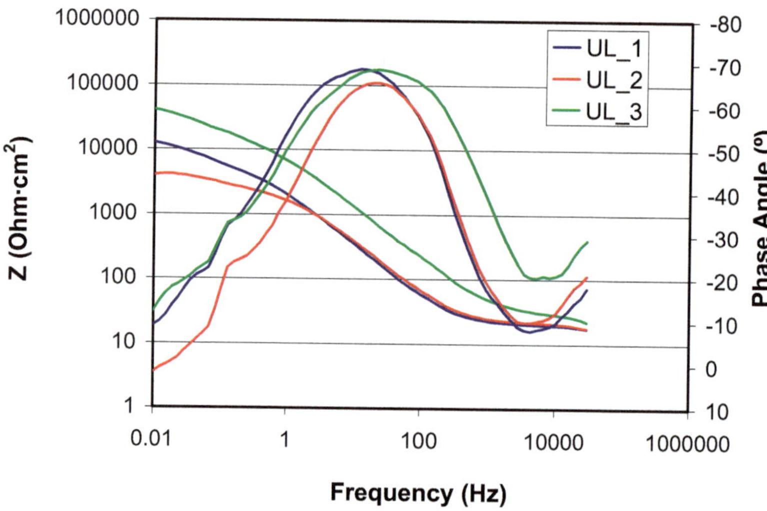
1000000
100000
10000
1000
100
10
1
Z (Ohm·cm²)
0.01
1
100
10000
1000000
Frequency (Hz)
-80
-70
-60
-50
-40
-30
-20
-10
0
10
Phase Angle (°)
UL_1
UL_2
UL_3

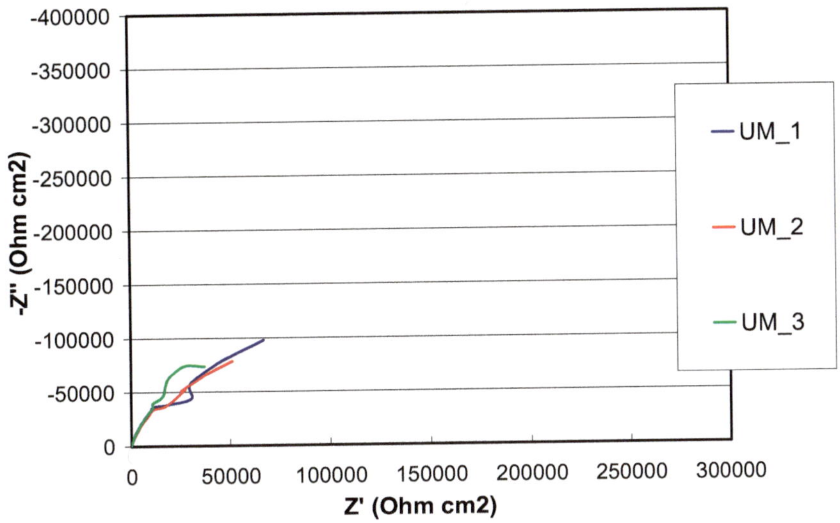
-400000
-350000
-300000
-250000
-200000
-150000
-100000
-50000
0
-Z'' (Ohm cm2)
0
50000
100000
150000
200000
250000
300000
Z' (Ohm cm2)
UM_1
UM_2
UM_3

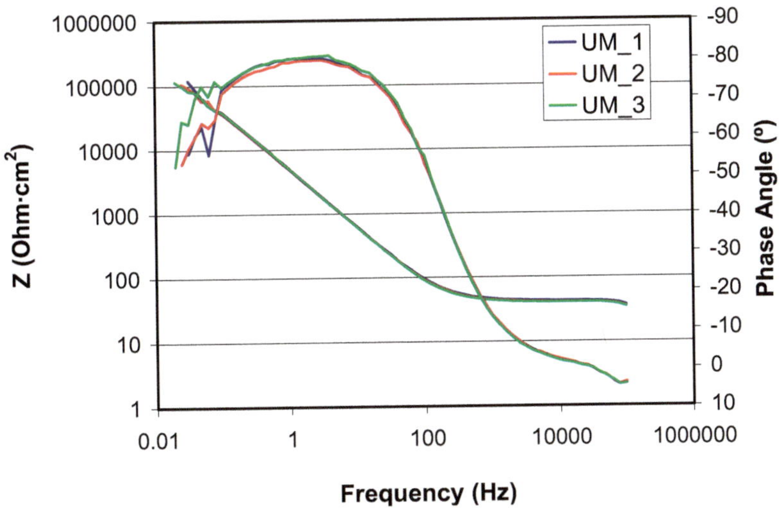
1000000
100000
10000
1000
100
10
1
Z (Ohm·cm2)
0.01
1
100
10000
1000000
Frequency (Hz)
-90
-80
-70
-60
-50
-40
-30
-20
-10
0
10
Phase Angle (º)
UM_1
UM_2
UM_3

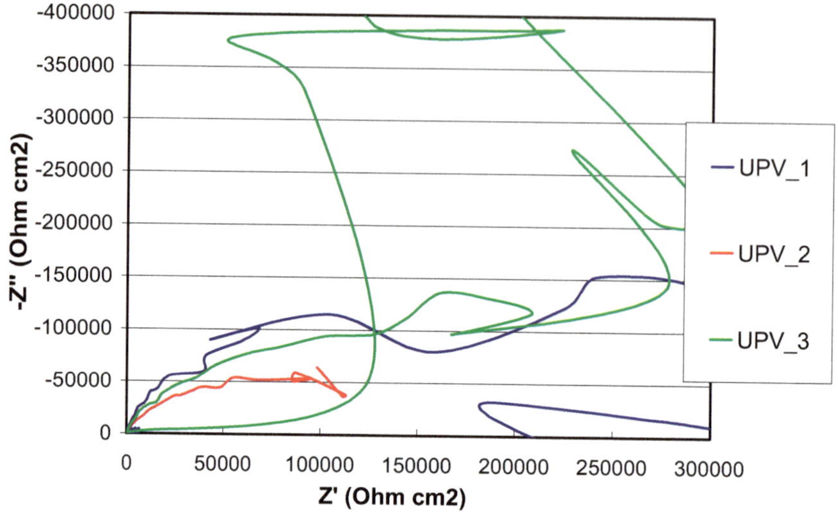

-Z'' (Ohm cm2)
-400000
-350000
-300000
-250000
-200000
-150000
-100000
-50000
0
0
50000
100000
150000
200000
250000
300000
Z' (Ohm cm2)
UPV_1
UPV_2
UPV_3

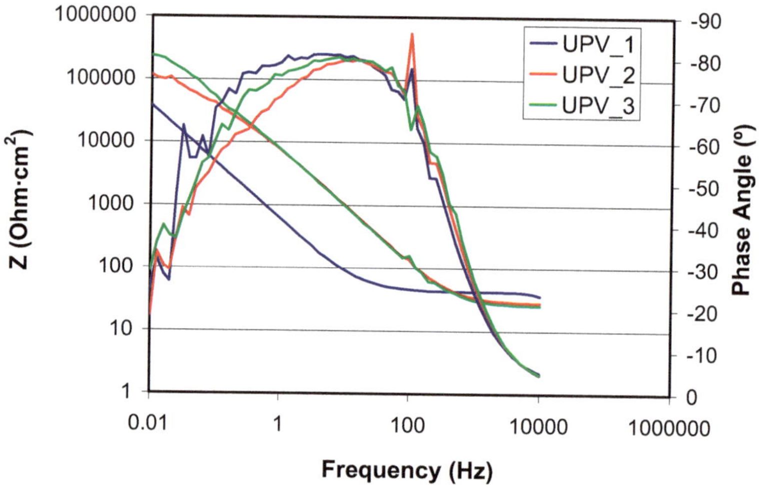

Z (Ohm·cm²)
1000000
100000
10000
1000
100
10
1
0.01
1
100
10000
1000000
Frequency (Hz)
Phase Angle (º)
-90
-80
-70
-60
-50
-40
-30
-20
-10
0
UPV_1
UPV_2
UPV_3

B.3 Impedance data **at 0.15 V_{SHE}** in PBS **with albumin**: 3 repeated tests for each laboratory

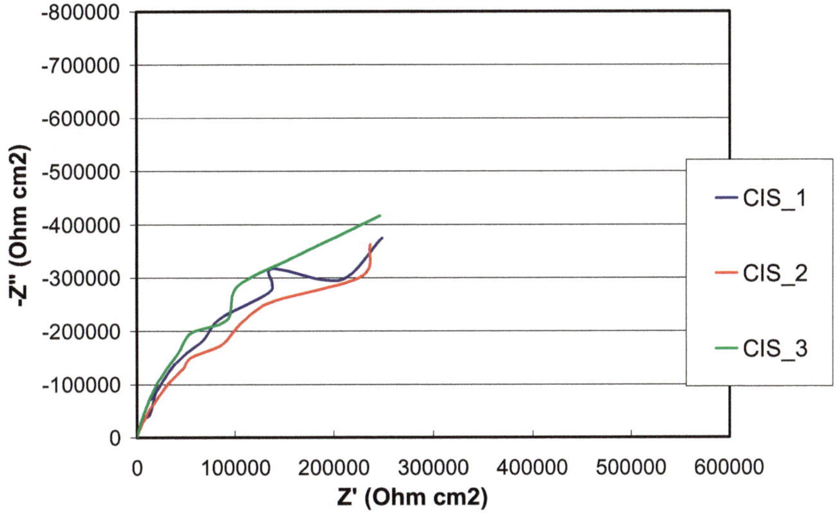

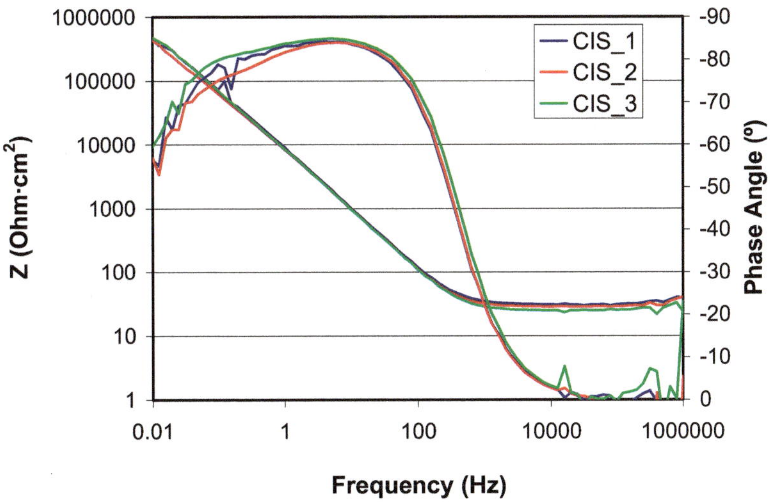

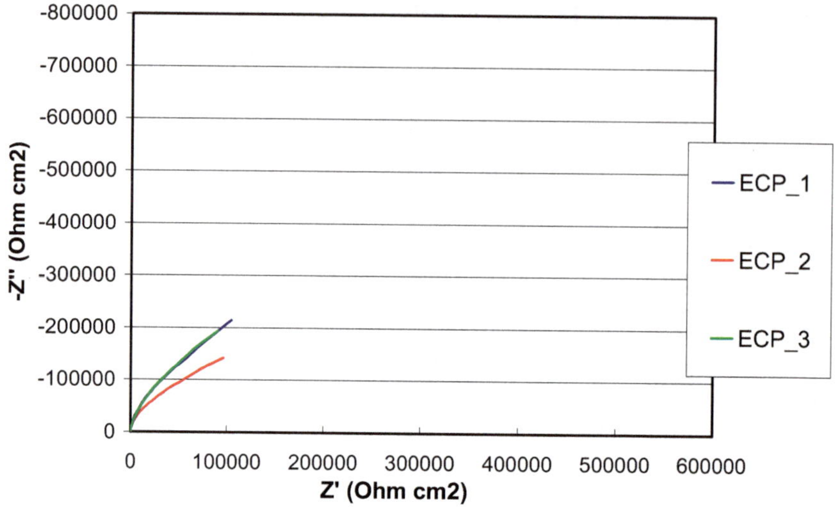
-800000
-700000
-600000
-500000
-400000
-300000
-200000
-100000
0
-Z'' (Ohm cm2)
0
100000
200000
300000
400000
500000
600000
Z' (Ohm cm2)
ECP_1
ECP_2
ECP_3

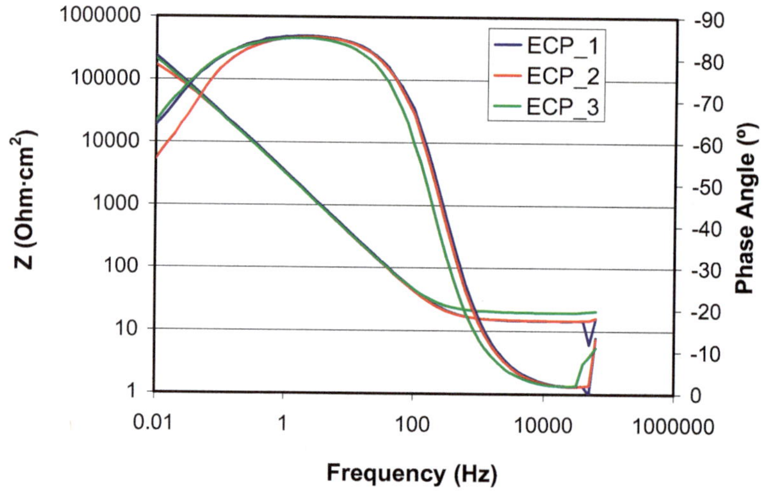
1000000
100000
10000
1000
100
10
1
Z (Ohm·cm²)
0.01
1
100
10000
1000000
Frequency (Hz)
-90
-80
-70
-60
-50
-40
-30
-20
-10
0
Phase Angle (º)
ECP_1
ECP_2
ECP_3

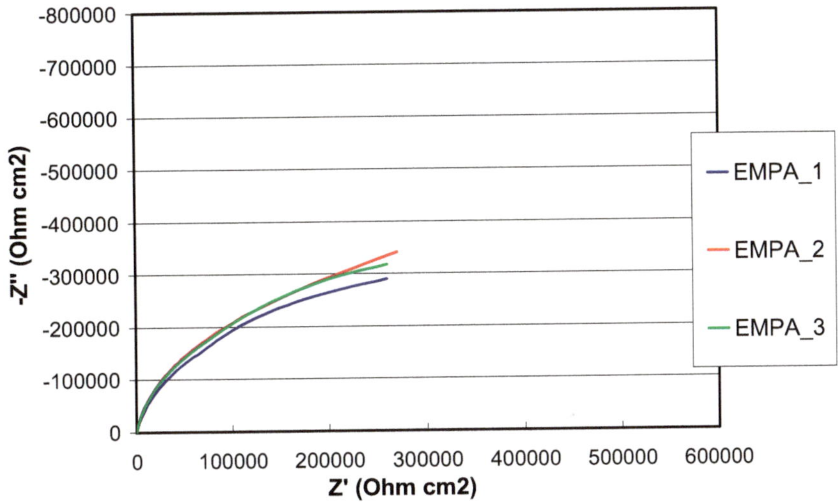
-800000
-700000
-600000
-500000
-400000
-300000
-200000
-100000
0
-Z'' (Ohm cm2)
0
100000
200000
300000
400000
500000
600000
Z' (Ohm cm2)
EMPA_1
EMPA_2
EMPA_3

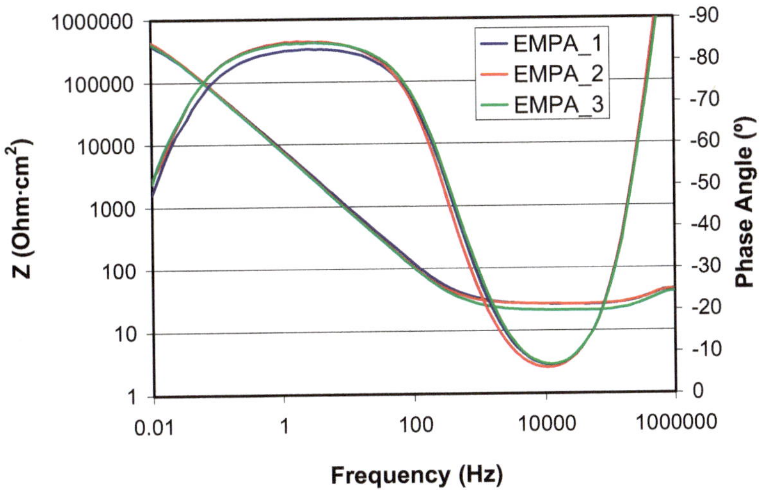
1000000
100000
10000
1000
100
10
1
Z (Ohm·cm²)
0.01
1
100
10000
1000000
Frequency (Hz)
-90
-80
-70
-60
-50
-40
-30
-20
-10
0
Phase Angle (º)
EMPA_1
EMPA_2
EMPA_3

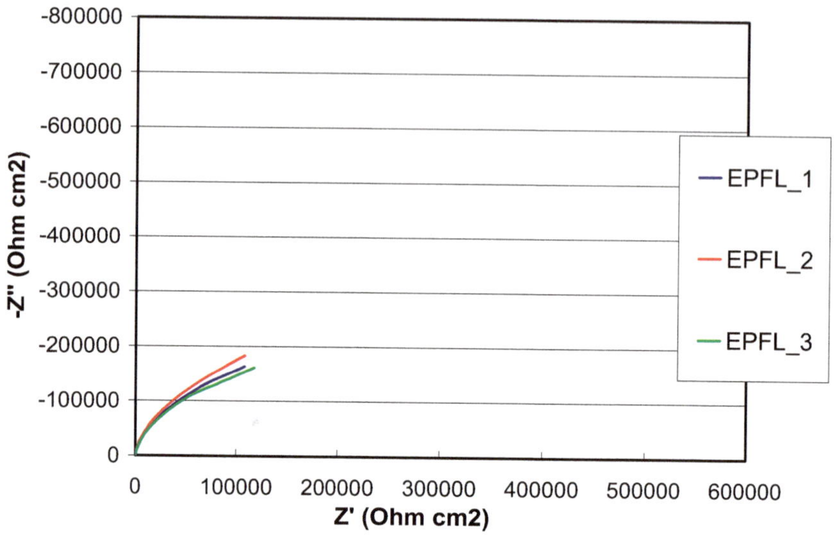
-800000
-700000
-600000
-500000
-400000
-300000
-200000
-100000
0
-Z" (Ohm cm2)
0
100000
200000
300000
400000
500000
600000
Z' (Ohm cm2)
EPFL_1
EPFL_2
EPFL_3

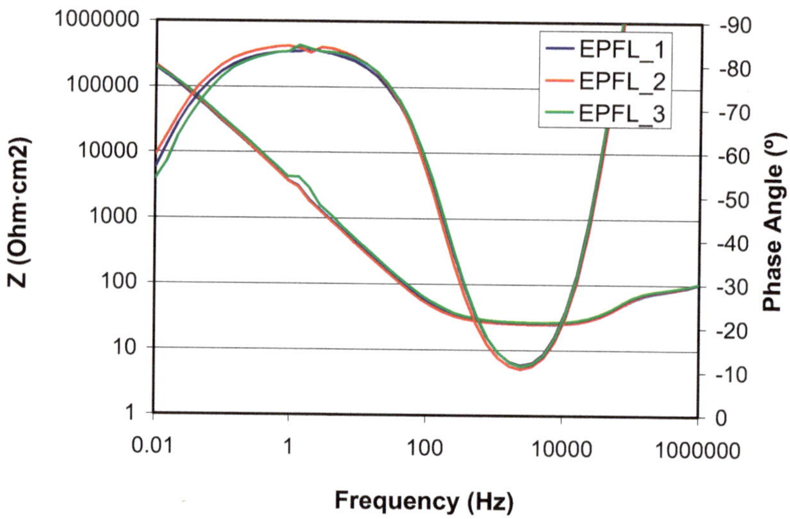
1000000
100000
10000
1000
100
10
1
Z (Ohm·cm2)
0.01
1
100
10000
1000000
Frequency (Hz)
-90
-80
-70
-60
-50
-40
-30
-20
-10
0
Phase Angle (°)
EPFL_1
EPFL_2
EPFL_3

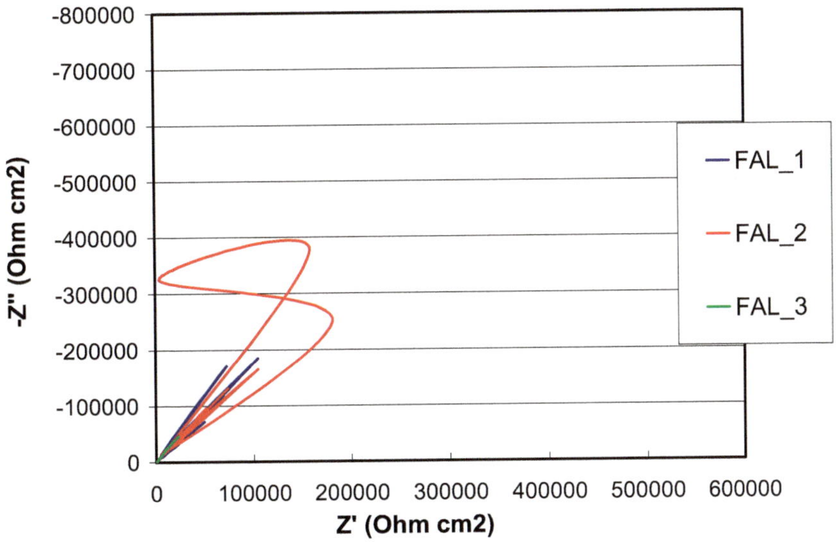
-800000
-700000
-600000
-500000
-400000
-300000
-200000
-100000
0
-Z'' (Ohm cm2)
0
100000
200000
300000
400000
500000
600000
Z' (Ohm cm2)
FAL_1
FAL_2
FAL_3

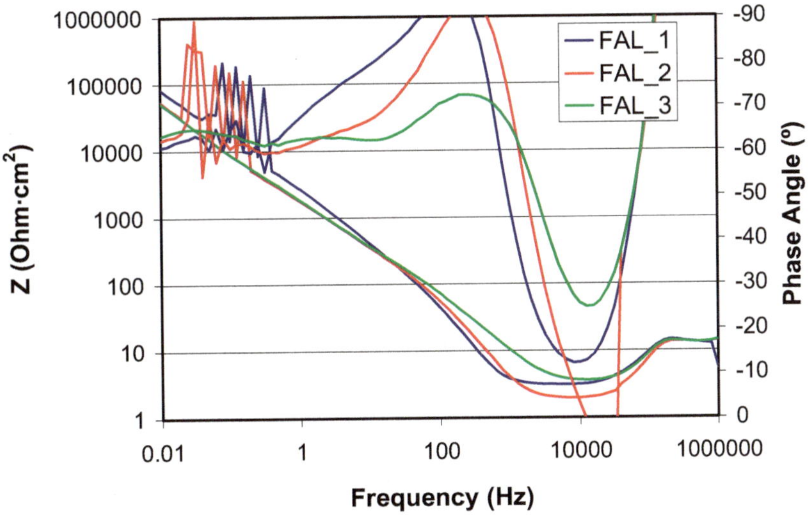
1000000
100000
10000
1000
100
10
1
Z (Ohm·cm2)
0.01
1
100
10000
1000000
Frequency (Hz)
-90
-80
-70
-60
-50
-40
-30
-20
-10
0
Phase Angle (º)
FAL_1
FAL_2
FAL_3

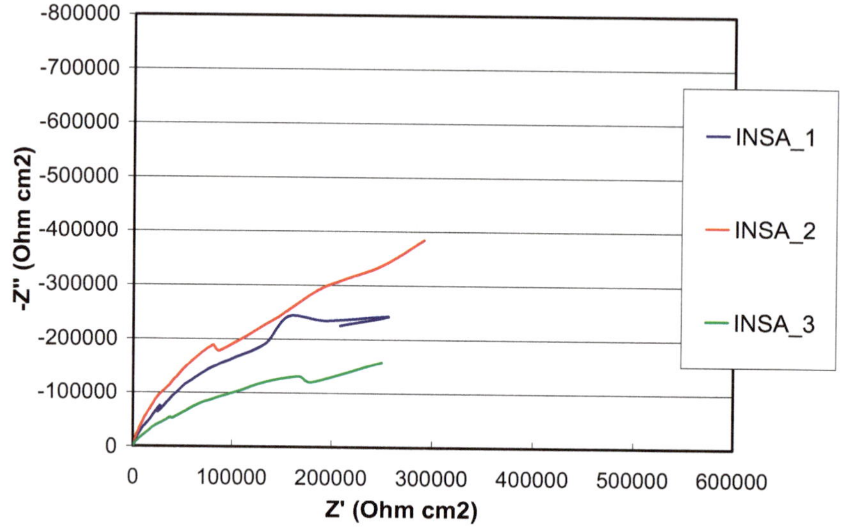
-800000
-700000
-600000
-500000
-400000
-300000
-200000
-100000
0
-Z'' (Ohm cm2)
0
100000
200000
300000
400000
500000
600000
Z' (Ohm cm2)
INSA_1
INSA_2
INSA_3

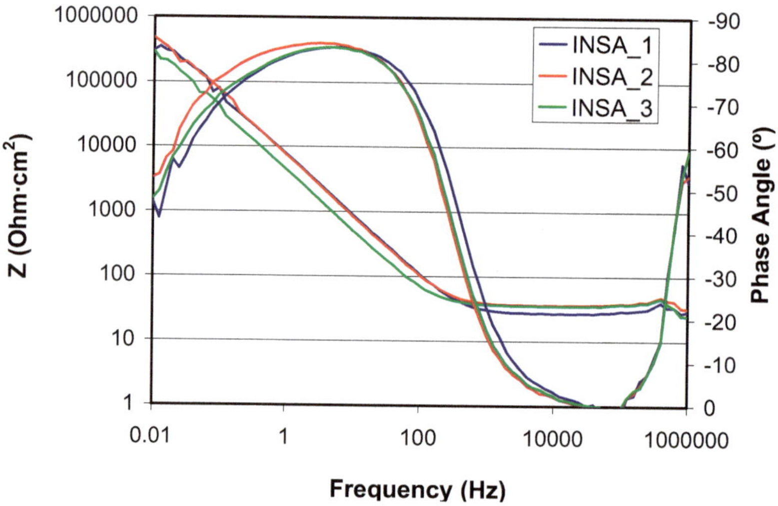
1000000
100000
10000
1000
100
10
1
Z (Ohm·cm^2)
0.01
1
100
10000
1000000
Frequency (Hz)
-90
-80
-70
-60
-50
-40
-30
-20
-10
0
Phase Angle (º)
INSA_1
INSA_2
INSA_3

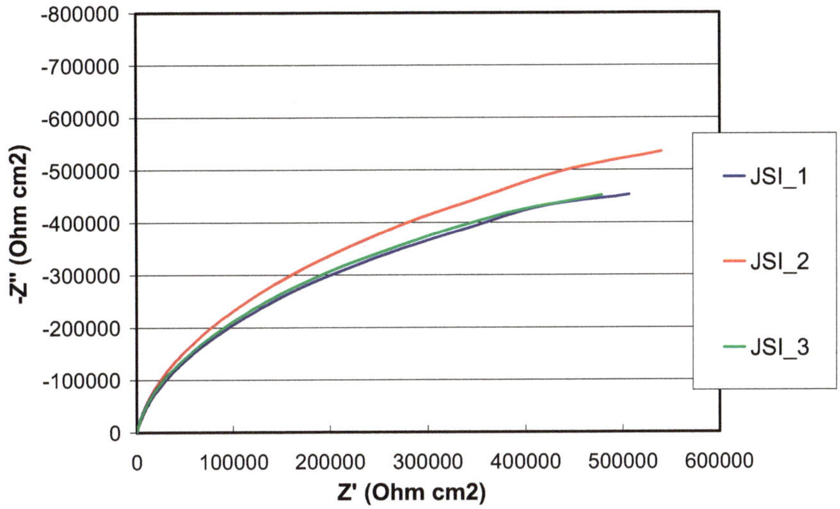
-800000
-700000
-600000
-500000
-400000
-300000
-200000
-100000
0
-Z" (Ohm cm2)
0
100000
200000
300000
400000
500000
600000
Z' (Ohm cm2)
JSI_1
JSI_2
JSI_3

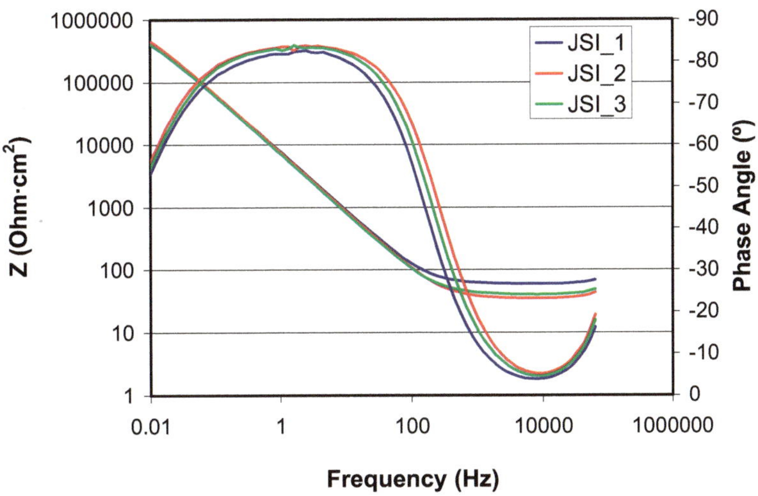
1000000
100000
10000
1000
100
10
1
Z (Ohm·cm²)
-90
-80
-70
-60
-50
-40
-30
-20
-10
0
Phase Angle (°)
0.01
1
100
10000
1000000
Frequency (Hz)
JSI_1
JSI_2
JSI_3

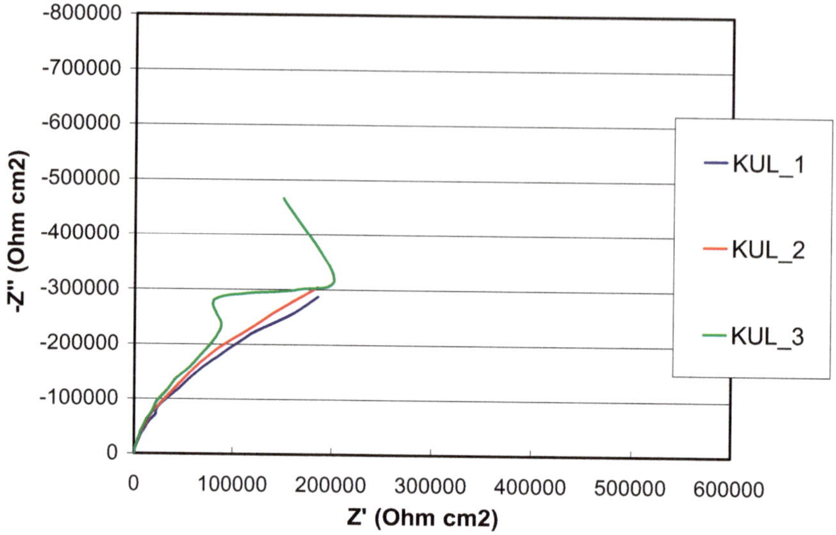
-800000
-700000
-600000
-500000
-400000
-300000
-200000
-100000
0
-Z'' (Ohm cm2)
0
100000
200000
300000
400000
500000
600000
Z' (Ohm cm2)
KUL_1
KUL_2
KUL_3

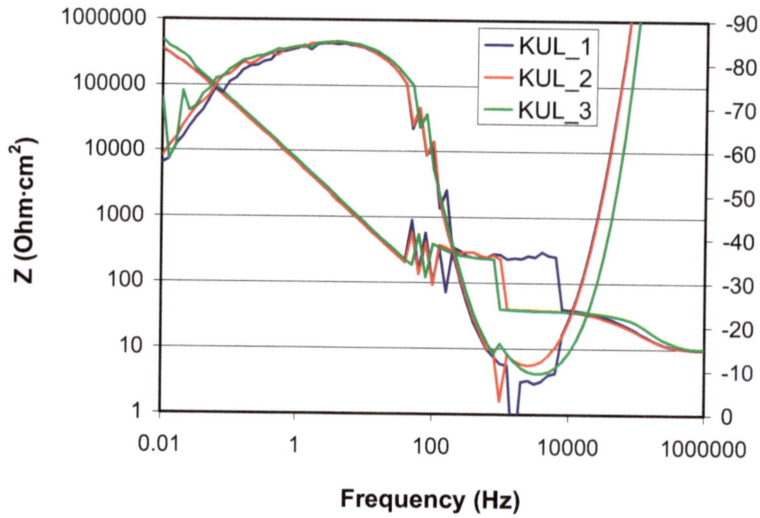
1000000
100000
10000
1000
100
10
1
Z (Ohm·cm2)
0.01
1
100
10000
1000000
Frequency (Hz)
-90
-80
-70
-60
-50
-40
-30
-20
-10
0
Phase Angle (°)
KUL_1
KUL_2
KUL_3

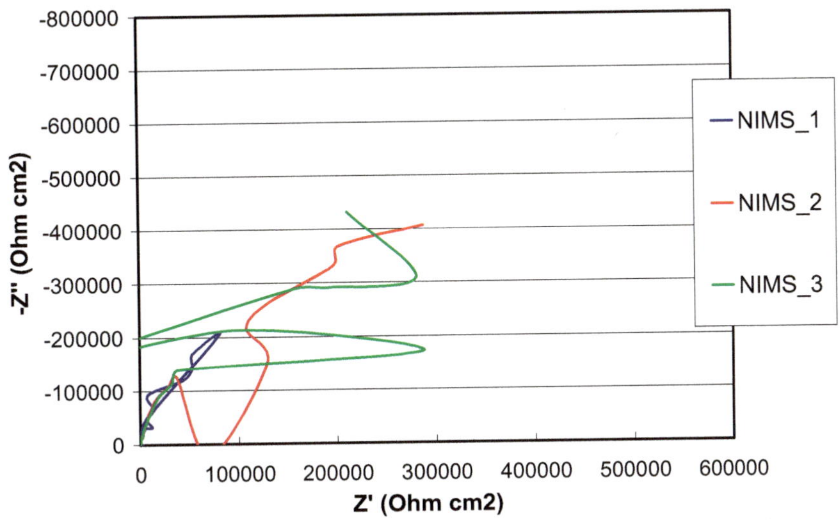
-800000
-700000
-600000
-500000
-400000
-300000
-200000
-100000
0
-Z'' (Ohm cm2)
0
100000
200000
300000
400000
500000
600000
Z' (Ohm cm2)
NIMS_1
NIMS_2
NIMS_3

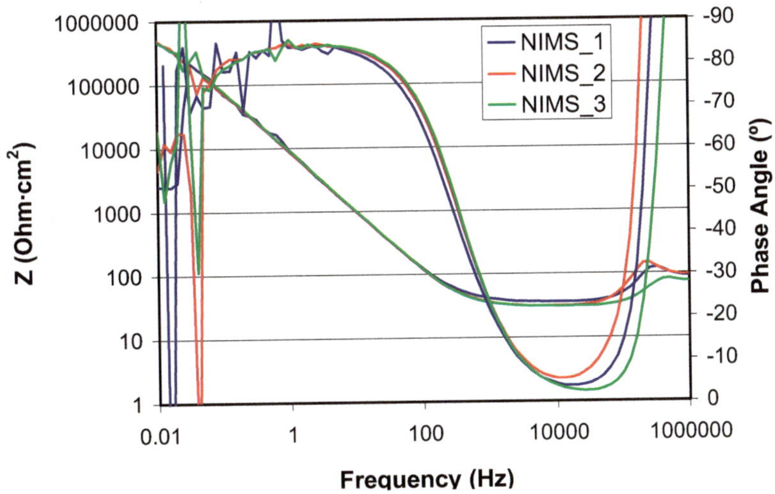
1000000
100000
10000
1000
100
10
1
Z (Ohm·cm²)
0.01
1
100
10000
1000000
Frequency (Hz)
-90
-80
-70
-60
-50
-40
-30
-20
-10
0
Phase Angle (°)
NIMS_1
NIMS_2
NIMS_3

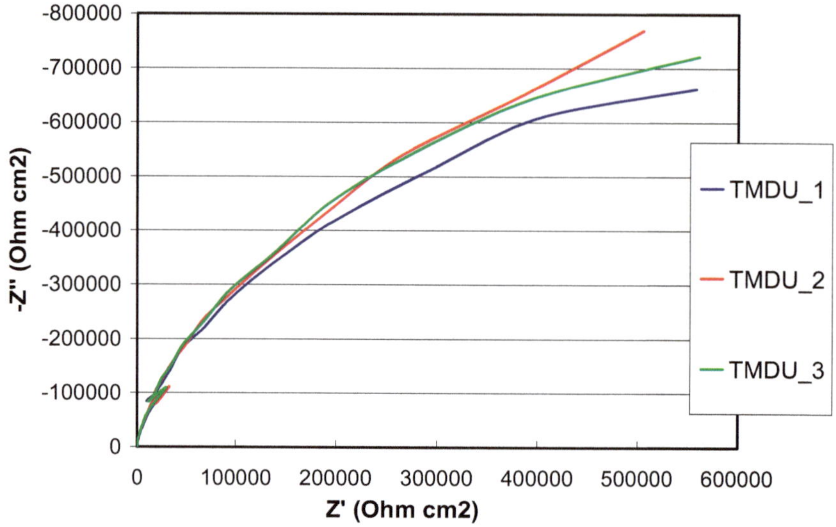
-800000
-700000
-600000
-500000
-400000
-300000
-200000
-100000
0
-Z'' (Ohm cm2)
0
100000
200000
300000
400000
500000
600000
Z' (Ohm cm2)
TMDU_1
TMDU_2
TMDU_3

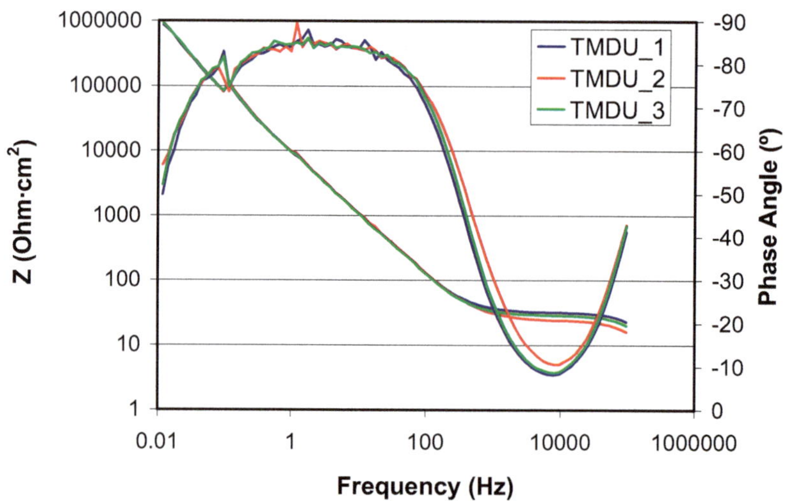
1000000
100000
10000
1000
100
10
1
Z (Ohm·cm^2)
-90
-80
-70
-60
-50
-40
-30
-20
-10
0
Phase Angle (°)
0.01
1
100
10000
1000000
Frequency (Hz)
TMDU_1
TMDU_2
TMDU_3

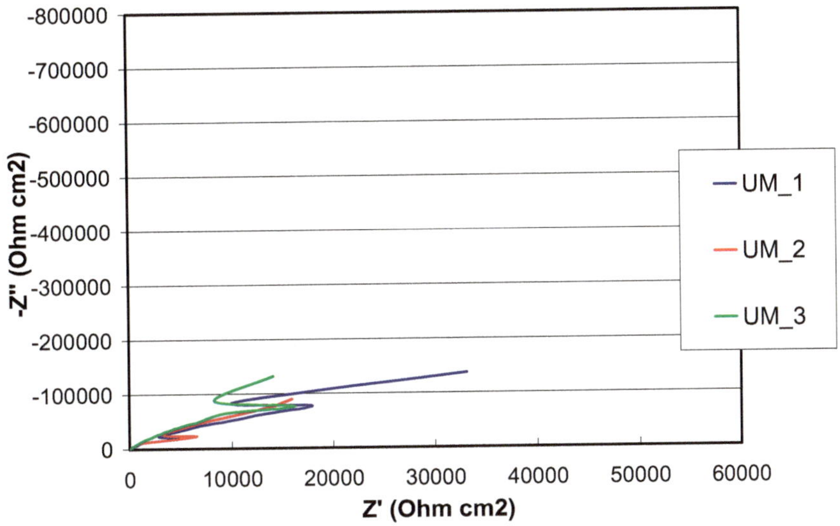
-800000
-700000
-600000
-500000
-400000
-300000
-200000
-100000
0
-Z'' (Ohm cm2)
0
10000
20000
30000
40000
50000
60000
Z' (Ohm cm2)
UM_1
UM_2
UM_3

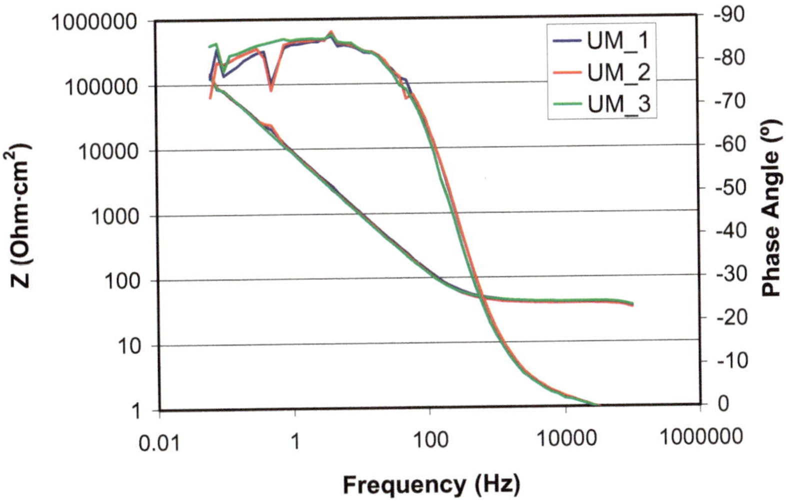
1000000
100000
10000
1000
100
10
1
Z (Ohm·cm^2)
0.01
1
100
10000
1000000
Frequency (Hz)
-90
-80
-70
-60
-50
-40
-30
-20
-10
0
Phase Angle (°)
UM_1
UM_2
UM_3

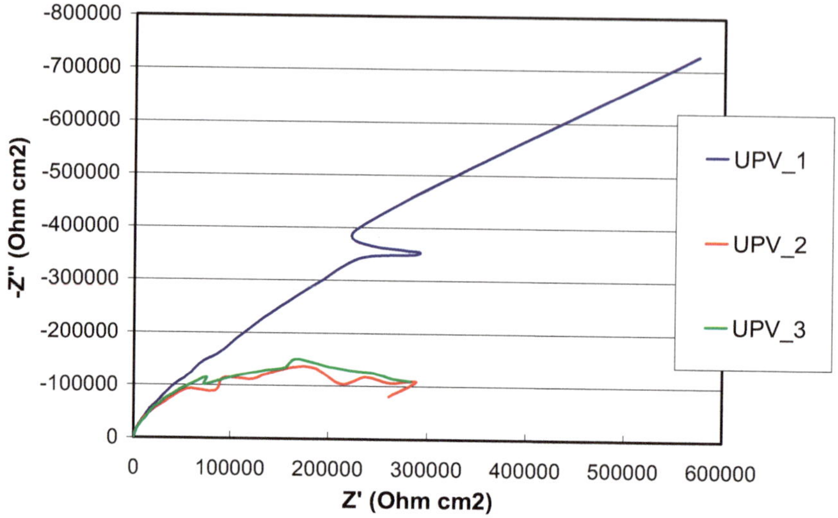
-800000
-700000
-600000
-500000
-400000
-300000
-200000
-100000
0
-Z'' (Ohm cm2)
0
100000
200000
300000
400000
500000
600000
Z' (Ohm cm2)
UPV_1
UPV_2
UPV_3

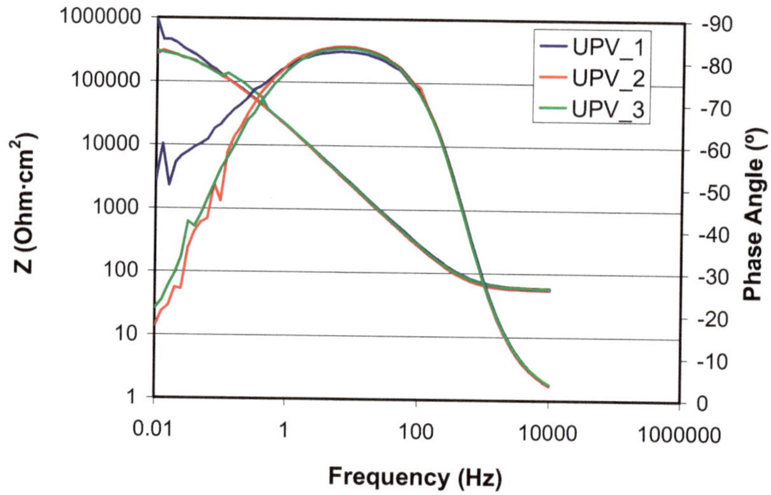
1000000
100000
10000
1000
100
10
1
Z (Ohm·cm²)
0.01
1
100
10000
1000000
Frequency (Hz)
-90
-80
-70
-60
-50
-40
-30
-20
-10
0
Phase Angle (º)
UPV_1
UPV_2
UPV_3

B.4 Impedance data **at OCP** in PBS **with albumin**: 3 repeated tests for each laboratory

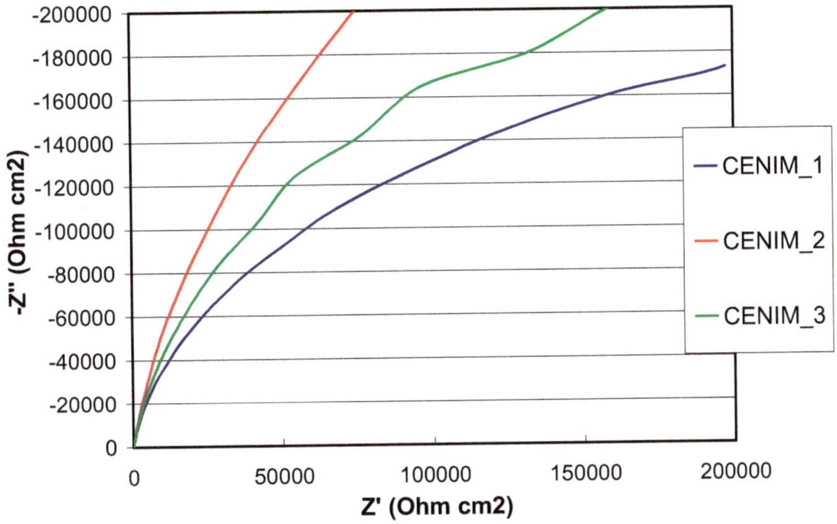

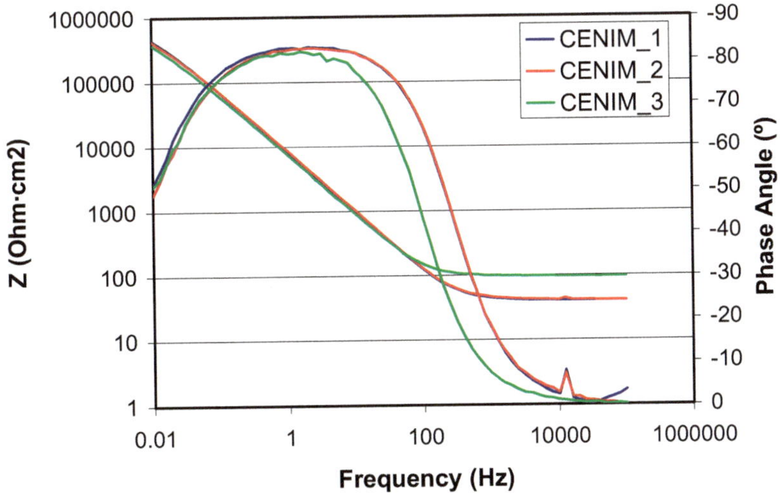

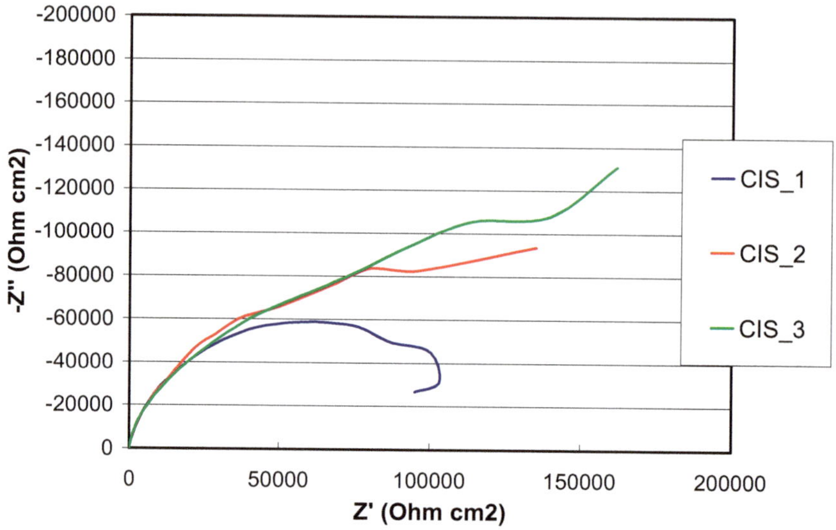

-200000
-180000
-160000
-140000
-120000
-100000
-80000
-60000
-40000
-20000
0
-Z'' (Ohm cm2)
0
50000
100000
150000
200000
Z' (Ohm cm2)
CIS_1
CIS_2
CIS_3

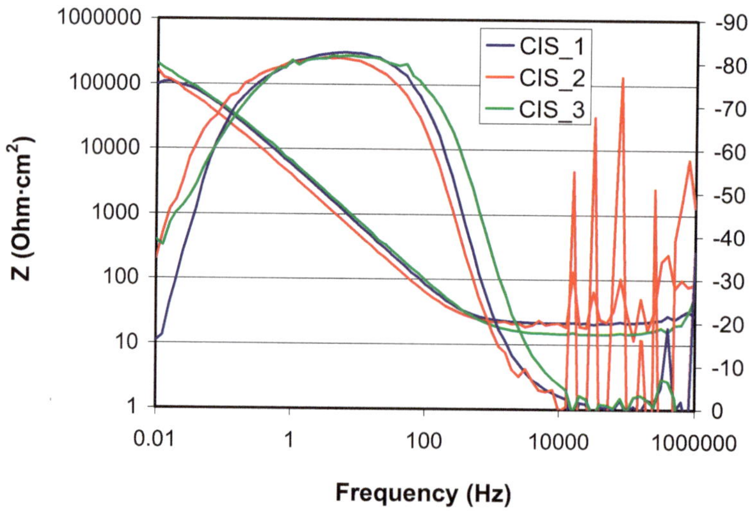

1000000
100000
10000
1000
100
10
1
Z (Ohm·cm²)
0.01
1
100
10000
1000000
Frequency (Hz)
-90
-80
-70
-60
-50
-40
-30
-20
-10
0
Phase Angle (°)
CIS_1
CIS_2
CIS_3

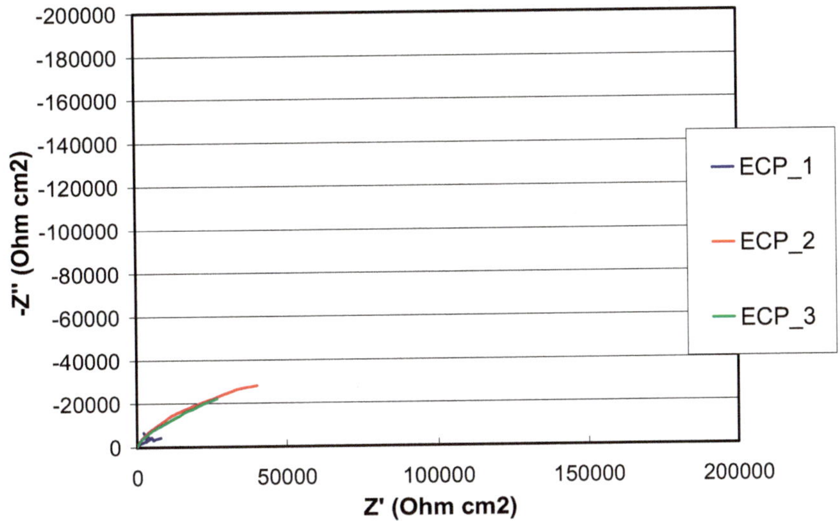
-200000
-180000
-160000
-140000
-120000
-100000
-80000
-60000
-40000
-20000
0
-Z" (Ohm cm2)
0
50000
100000
150000
200000
Z' (Ohm cm2)
ECP_1
ECP_2
ECP_3

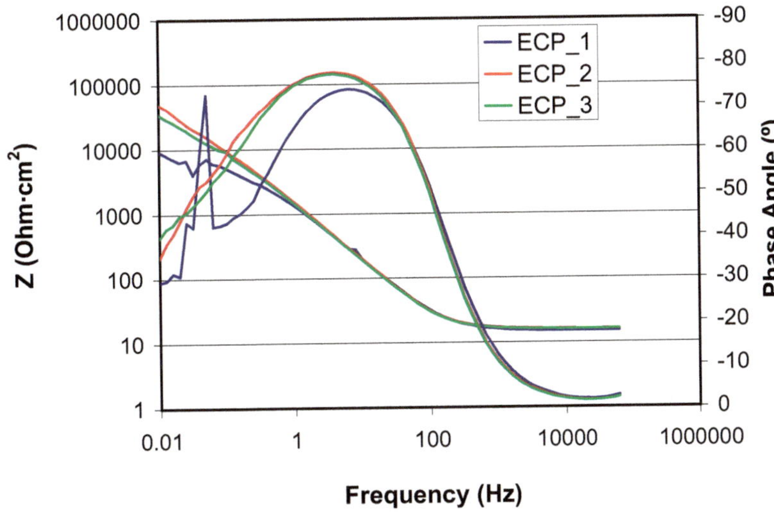
1000000
100000
10000
1000
100
10
1
Z (Ohm·cm²)
0.01
1
100
10000
1000000
Frequency (Hz)
-90
-80
-70
-60
-50
-40
-30
-20
-10
0
Phase Angle (º)
ECP_1
ECP_2
ECP_3

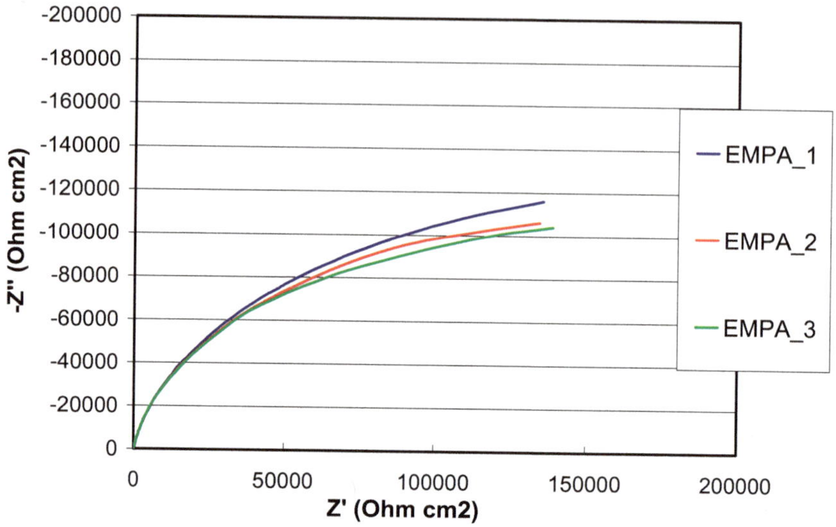
-200000
-180000
-160000
-140000
-120000
-100000
-80000
-60000
-40000
-20000
0
-Z" (Ohm cm2)
0
50000
100000
150000
200000
Z' (Ohm cm2)
EMPA_1
EMPA_2
EMPA_3

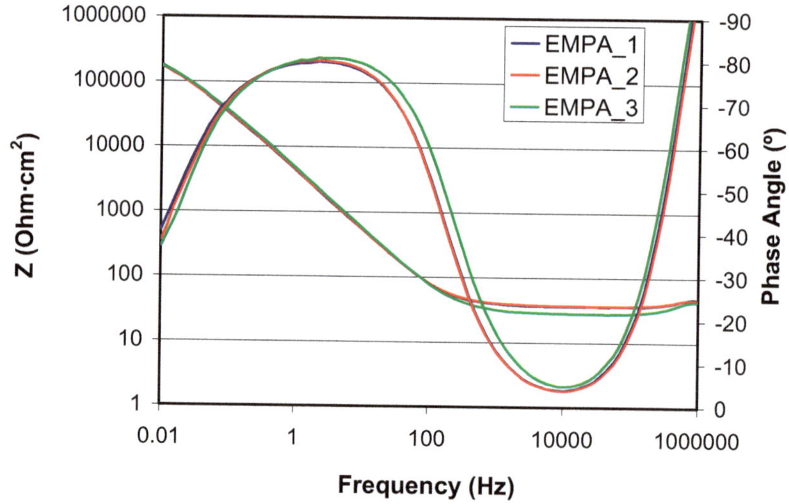
1000000
100000
10000
1000
100
10
1
Z (Ohm·cm²)
0.01
1
100
10000
1000000
Frequency (Hz)
-90
-80
-70
-60
-50
-40
-30
-20
-10
0
Phase Angle (°)
EMPA_1
EMPA_2
EMPA_3

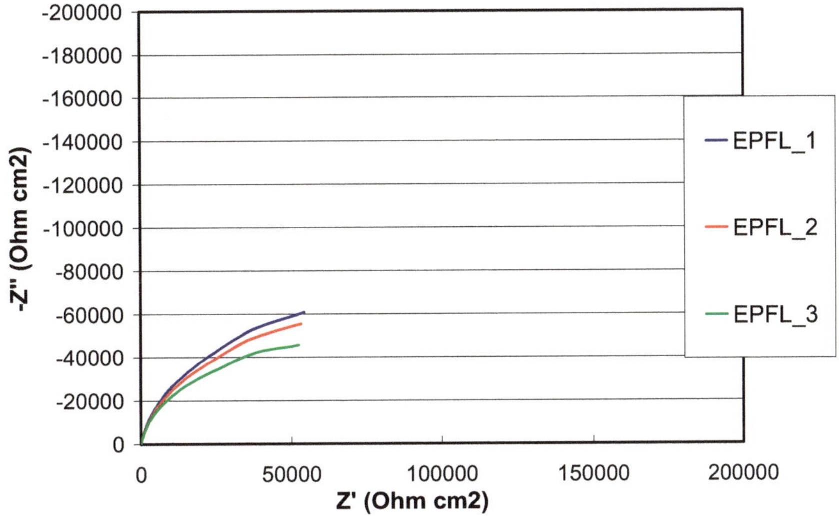
-200000
-180000
-160000
-140000
-120000
-100000
-80000
-60000
-40000
-20000
0
-Z" (Ohm cm2)
0
50000
100000
150000
200000
Z' (Ohm cm2)
EPFL_1
EPFL_2
EPFL_3

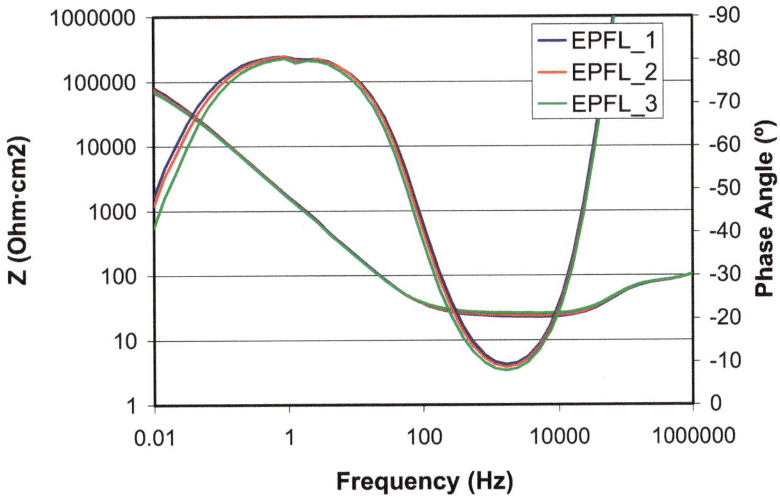
1000000
100000
10000
1000
100
10
1
Z (Ohm·cm2)
0.01
1
100
10000
1000000
Frequency (Hz)
-90
-80
-70
-60
-50
-40
-30
-20
-10
0
Phase Angle (º)
EPFL_1
EPFL_2
EPFL_3

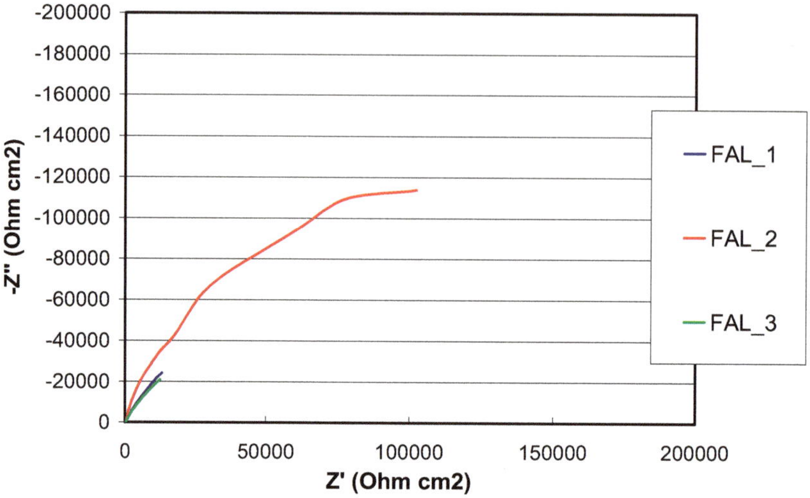
-Z" (Ohm cm2)
-200000
-180000
-160000
-140000
-120000
-100000
-80000
-60000
-40000
-20000
0
0
50000
100000
150000
200000
Z' (Ohm cm2)
FAL_1
FAL_2
FAL_3

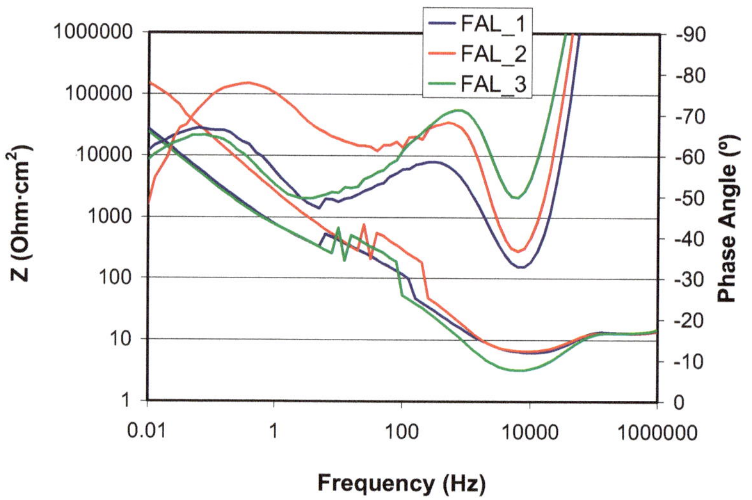
FAL_1
FAL_2
FAL_3
Z (Ohm·cm²)
1000000
100000
10000
1000
100
10
1
-90
-80
-70
-60
-50
-40
-30
-20
-10
0
Phase Angle (º)
0.01
1
100
10000
1000000
Frequency (Hz)

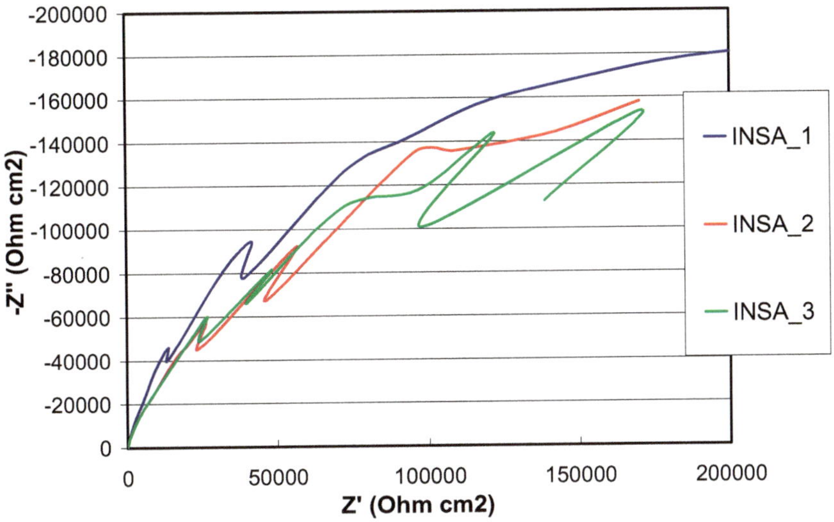
-200000
-180000
-160000
-140000
-120000
-100000
-80000
-60000
-40000
-20000
0
-Z'' (Ohm cm2)
0
50000
100000
150000
200000
Z' (Ohm cm2)
INSA_1
INSA_2
INSA_3

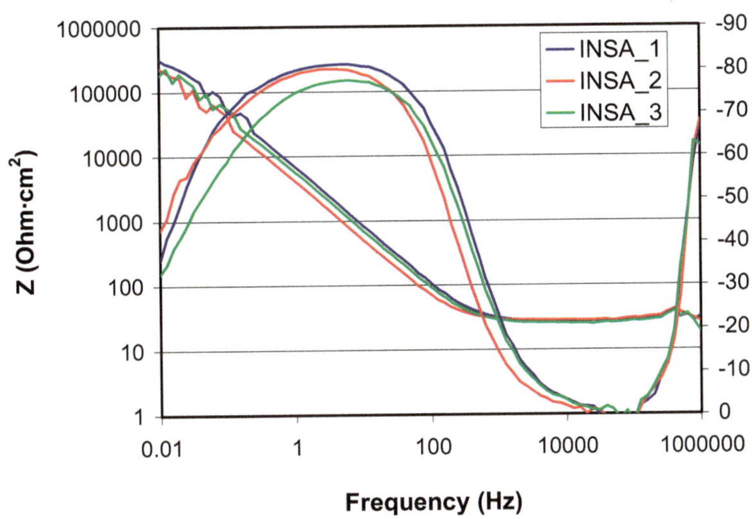
1000000
100000
10000
1000
100
10
1
Z (Ohm·cm2)
0.01
1
100
10000
1000000
Frequency (Hz)
-90
-80
-70
-60
-50
-40
-30
-20
-10
0
Phase Angle (°)
INSA_1
INSA_2
INSA_3

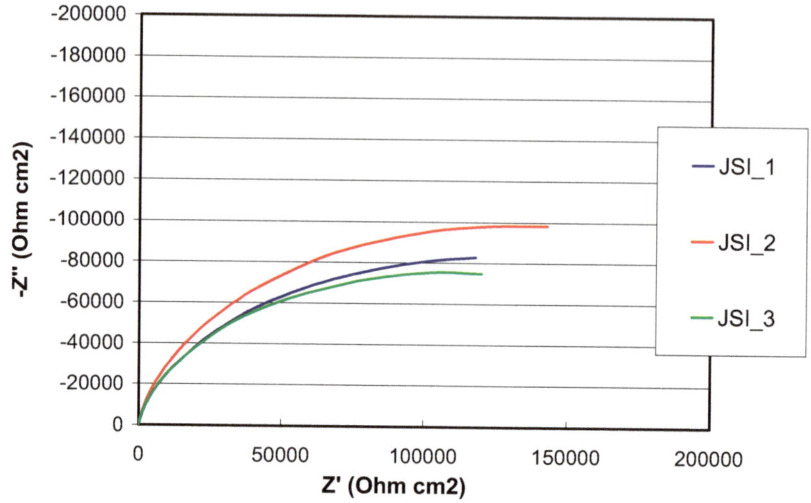
-200000
-180000
-160000
-140000
-120000
-100000
-80000
-60000
-40000
-20000
0
-Z'' (Ohm cm2)
0
50000
100000
150000
200000
Z' (Ohm cm2)
JSI_1
JSI_2
JSI_3

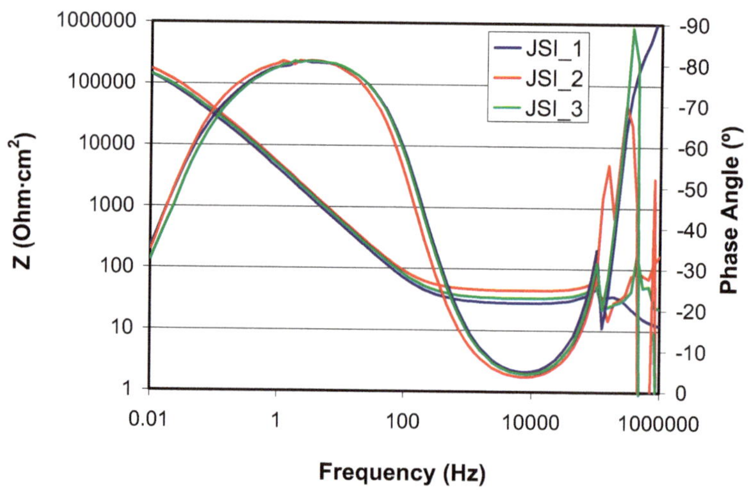
1000000
100000
10000
1000
100
10
1
Z (Ohm·cm^2)
0.01
1
100
10000
1000000
Frequency (Hz)
-90
-80
-70
-60
-50
-40
-30
-20
-10
0
Phase Angle (º)
JSI_1
JSI_2
JSI_3

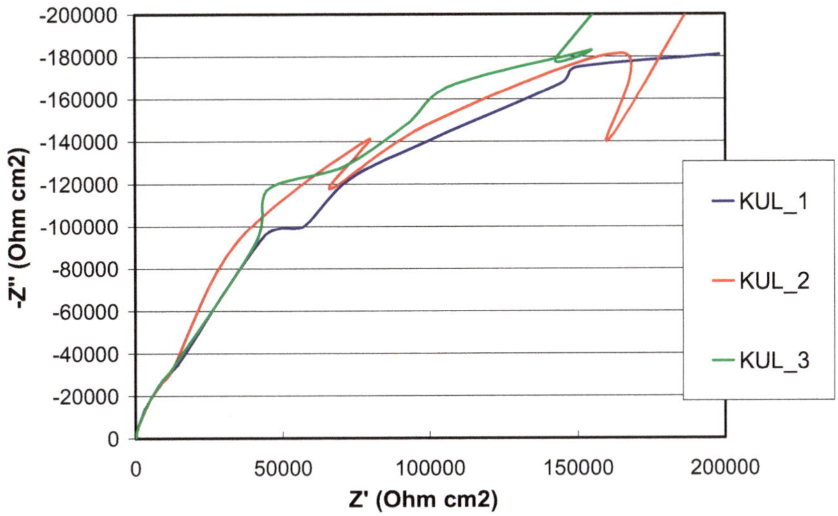
-200000
-180000
-160000
-140000
-120000
-100000
-80000
-60000
-40000
-20000
0
-Z'' (Ohm cm2)
0
50000
100000
150000
200000
Z' (Ohm cm2)
KUL_1
KUL_2
KUL_3

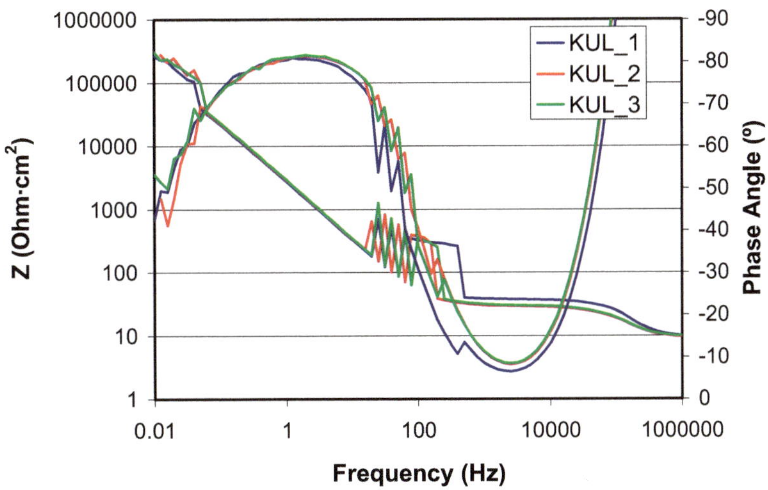
1000000
100000
10000
1000
100
10
1
Z (Ohm·cm^2)
0.01
1
100
10000
1000000
Frequency (Hz)
-90
-80
-70
-60
-50
-40
-30
-20
-10
0
Phase Angle (°)
KUL_1
KUL_2
KUL_3

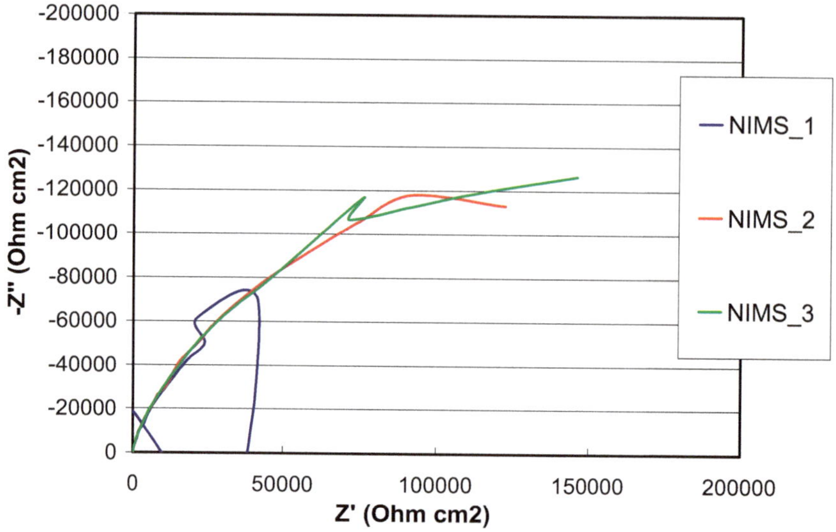
-200000
-180000
-160000
-140000
-120000
-100000
-80000
-60000
-40000
-20000
0
-Z'' (Ohm cm2)
0
50000
100000
150000
200000
Z' (Ohm cm2)
NIMS_1
NIMS_2
NIMS_3

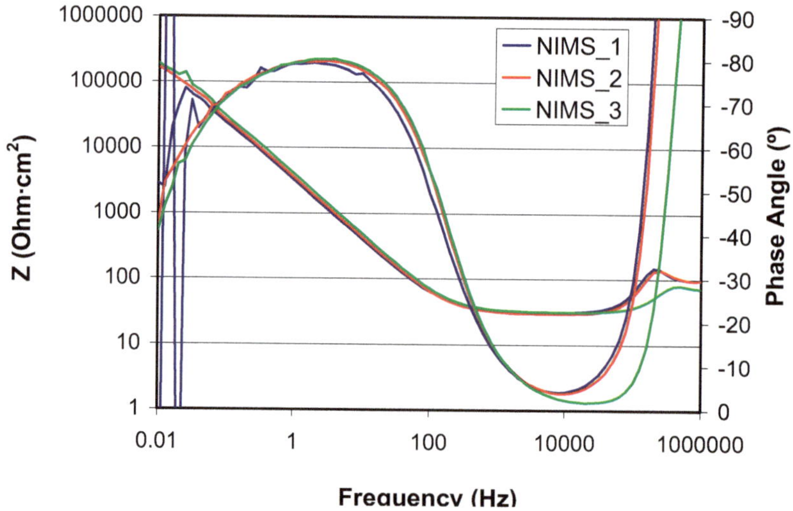
1000000
100000
10000
1000
100
10
1
Z (Ohm·cm^2)
-90
-80
-70
-60
-50
-40
-30
-20
-10
0
Phase Angle (°)
0.01
1
100
10000
1000000
Frequency (Hz)
NIMS_1
NIMS_2
NIMS_3

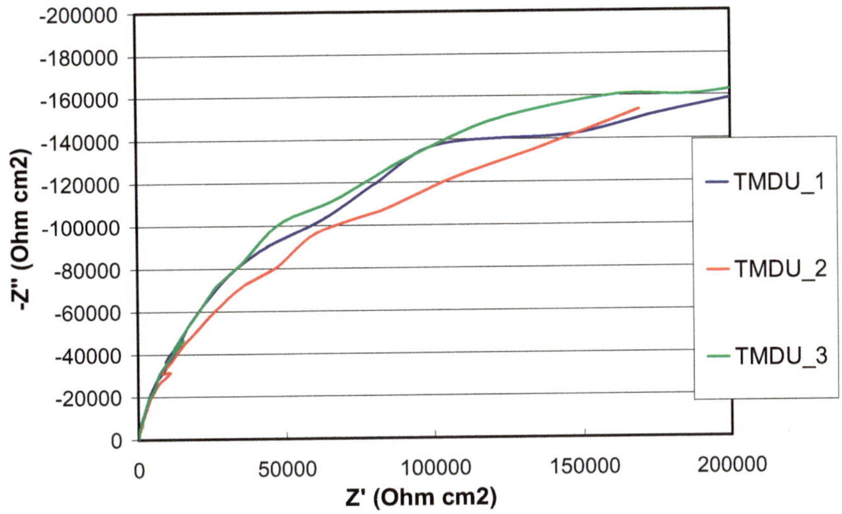
-200000
-180000
-160000
-140000
-120000
-100000
-80000
-60000
-40000
-20000
0
-Z" (Ohm cm2)
0
50000
100000
150000
200000
Z' (Ohm cm2)
TMDU_1
TMDU_2
TMDU_3

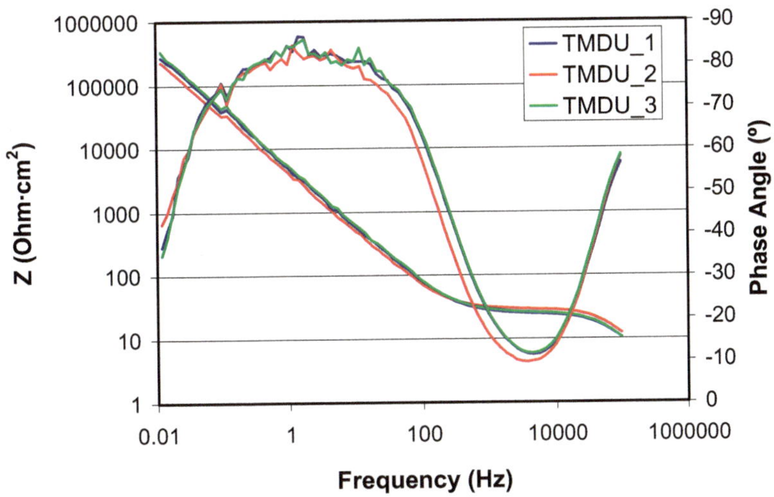
1000000
100000
10000
1000
100
10
1
Z (Ohm·cm^2)
0.01
1
100
10000
1000000
Frequency (Hz)
-90
-80
-70
-60
-50
-40
-30
-20
-10
0
Phase Angle (°)
TMDU_1
TMDU_2
TMDU_3

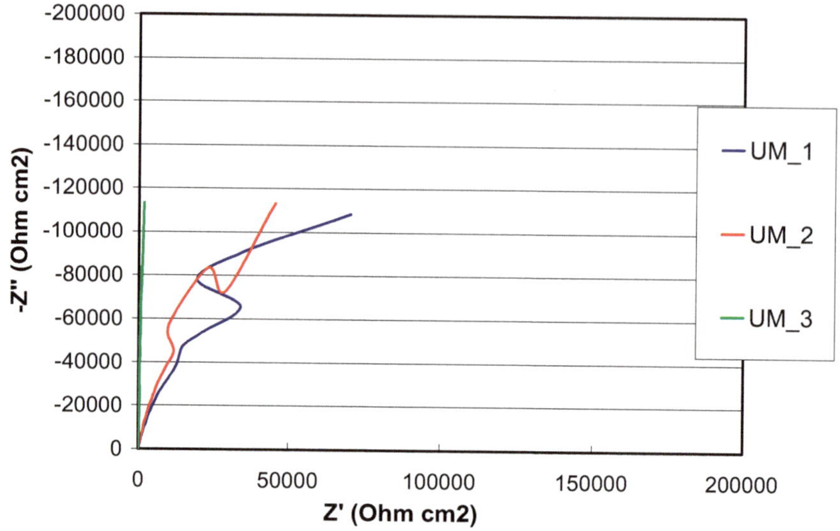
-200000
-180000
-160000
-140000
-120000
-100000
-80000
-60000
-40000
-20000
0
-Z'' (Ohm cm2)
0
50000
100000
150000
200000
Z' (Ohm cm2)
UM_1
UM_2
UM_3

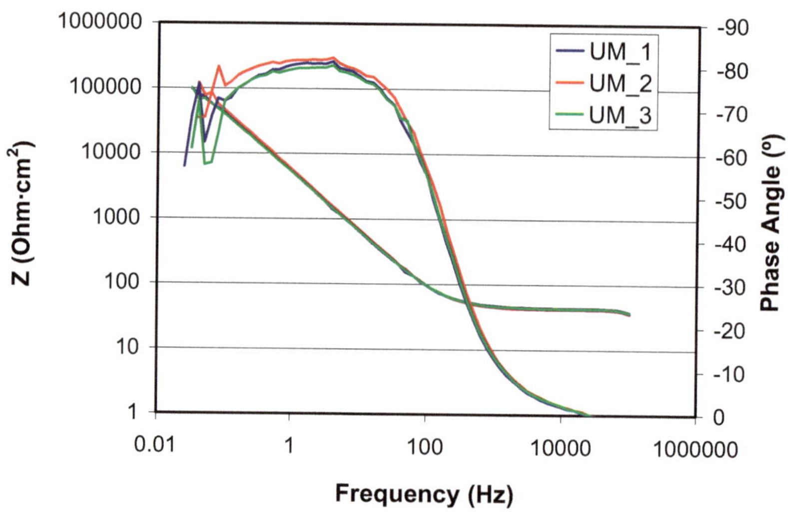
1000000
100000
10000
1000
100
10
1
Z (Ohm·cm^2)
-90
-80
-70
-60
-50
-40
-30
-20
-10
0
Phase Angle (°)
0.01
1
100
10000
1000000
Frequency (Hz)
UM_1
UM_2
UM_3

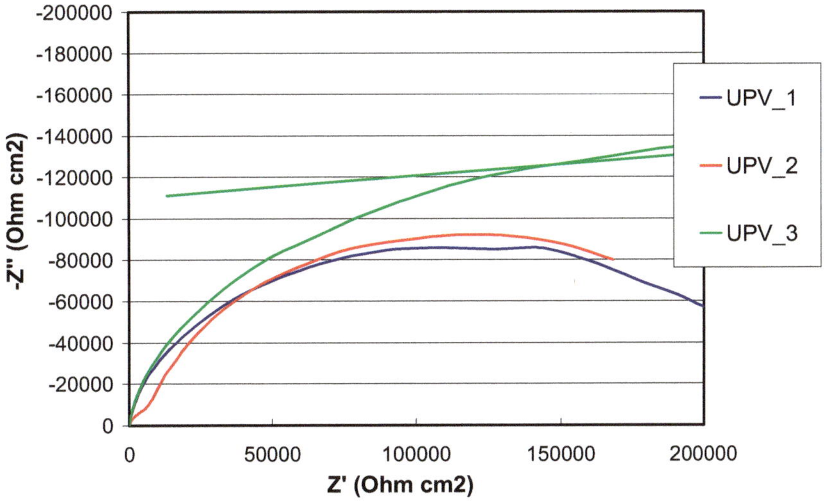
-200000
-180000
-160000
-140000
-120000
-100000
-80000
-60000
-40000
-20000
0
-Z'' (Ohm cm2)
0
50000
100000
150000
200000
Z' (Ohm cm2)
UPV_1
UPV_2
UPV_3

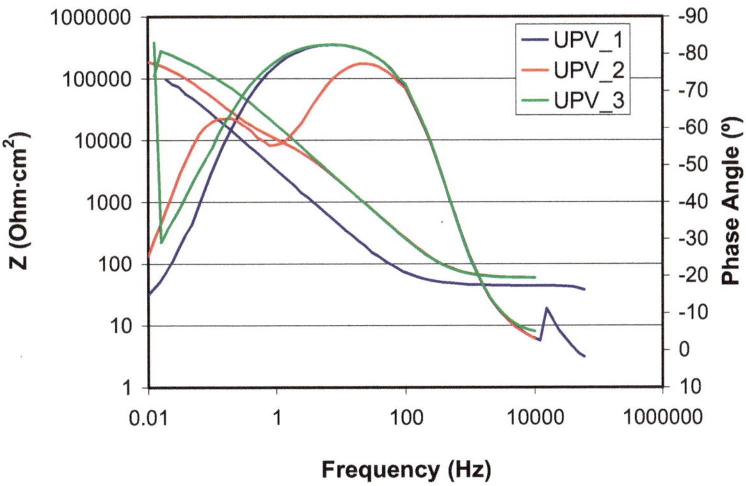
1000000
100000
10000
1000
100
10
1
Z (Ohm·cm²)
0.01
1
100
10000
1000000
Frequency (Hz)
-90
-80
-70
-60
-50
-40
-30
-20
-10
0
10
Phase Angle (°)
UPV_1
UPV_2
UPV_3

References

1. S. Virtanen, I. Milosev, E. Gomerz-Barrena, R. Trebse, J. Salo and Y. T. Konttinen, *Acta Biomater.*, 4 (2008), 468–476.
2. J. Black, *J. Bone Joint Surg.*, 70B (1988), 517–520.
3. H. Luthy, C. P. Marinello, L. Reclaru and P. Scharer, *J. Prosthet. Dent.*, 75(5) (1996), 515–524.
4. J. R. Goldberg and J. L. Gilbert, *J. Biomed. Mater. Res.*, (1997).
5. H. C. Hsu and S. K. Yen, *Dent. Mater.*, 14 (1998), 339–346.
6. F. Contu, B. Elsener and H. Böhni, *J. Biomed. Mater. Res.*, 62A (2002), 412–421.
7. F. Contu, B. Elesener and H. Bohni, *J. Biomed. Mater. Res.*, 67A (2003), 246–254.
8. F. Contu, B. Elesener and H. Bohni, *J. Electrochem. Soc.*, 150(9) (2003), B419–B424.
9. I. Milosev and H.-H. Strehblow, *Electrochim. Acta*, 48 (2003), 2767–2774.
10. A. Dorner-Reisel, C. Schürer, G. Irmer and E. Müller, *Surf. Coat. Technol.*, 177–178 (2004), 830–837.
11. A. W. E. Hodgson, S. Kurz, S. Virtanen, V. Fervel, C. O. A. Olsson and S. Mischler, *Electrochim. Acta*, 49 (2004), 2167–2178.
12. A. Kocijan, I. Milosev, D. K. Merl and B. Pihlar, *J. Appl. Electrochem.*, 34 (2004), 517–524.
13. F. Contu, B. Elsener and H. Böhni, *Corros. Sci.*, 47 (2005), 1863–1875.
14. S. Hiromoto, E. Onodera, A. Chiba, K. Asami and T. Hanawa, *Biomaterials*, 26 (2005), 4912–4923.
15. A. C. Lewis, M. R. Kilburn, I. Papageorgiou, G. C. Allen and C. P. Case, *J. Biomed. Mater. Res.*, 73(4) (2005), 456–467.
16. L. Reclaru, P. Y. Eschler, R. Lerf and A. Blatter, *Biomaterials*, 26 (2005), 4747–4756.
17. M. Metikos-Hukovic, Z. Pilic, R. Babic and D. Omanovic, *Acta Biomater.*, 2(6) (2006), 693–700.
18. O. Öztürk, U. Türkan and A. E. Eroglu, *Surf. Coat. Technol.*, 200 (2006), 5687–5697.
19. J. Baszkiewicz, D. Krupa, B. Rajchel, J. A. Kozubowski, A. Barcz, J. W. Sobczak *et al.*, *Vacuum*, 81 (2007), 1306–1309.
20. A. Igual-Muñoz and S. Mischler, *J. Electrochem. Soc.*, 154(10) (2007), C562–C570.
21. M. Metikos-Hukovic and R. Babic, *Corros. Sci.*, 49 (2007), 3570–3579.
22. Y. Yan, A. Neville and D. Dowson, *Wear*, 263(7–12) (2007), 1105–1111.
23. M. Metikos-Hukovic and R. Babic, *Corros. Sci.*, 51(1) (2009), 70–75.
24. Y. Okazaki and E. Gotoh, *Corros. Sci.*, 50 (2008), 3429–3438.
25. A. Ouerd, C. Alemany-Dumont, B. Normand and S. Szunerits, *Electrochim. Acta*, 53 (2008), 4461–4469.
26. C. Valero Vidal and A. Igual-Muñoz, *Corros. Sci.*, 50 (2008), 1954–1961.
27. C. Valero Vidal and A. Igual-Muñoz, *Electrochim. Acta*, 54 (2009), 1798–1809.
28. D. Sun, J. A. Wharton, R. J. K. Wood, L. Ma and W. M. Rainforth, *Tribol. Int.*, 42 (2009), 99–110.
29. B. A. Boukamp, *Solid State Ionics*, 18/19 (1986), 136–140.
30. J. Cawley, J. E. P. Metcalf, A. H. Jones, T. J. Band and D. S. Skupien, *Wear*, 255 (2003), 999–1006.
31. S. Dobbs and J. L. M. Robertson, *J. Mater. Sci.*, 18 (1983), 391–401.
32. H. E. Placko, S. A. Brown and J. H. Payer, *J. Biomed. Mater. Res.*, 39 (1998), 292–299.
33. S. Mischler and P. Ponthiaux, *Wear*, 248(211) (2001), 225.
34. P. Marcus and I. Olefjord, *Corros. Sci.*, 28(6) (1988), 589–602.